W9-BTD-352

To Buckminster Fuller
    friend of the universe
    bringer of happiness.
    liberator.

        With affectionate admiration

            Ezra Pound

        Spoleto
            June 29th 1971

    To Ron Wise
        friend of mine,
                        mine

Other books by R. Buckminster Fuller:

4 D Timelock
Nine Chains to the Moon
The Dymaxion World of Buckminster Fuller: with Robert Marks
Education Automation
Untitled Epic Poem on the History of Industrialization
Ideas and Integrities
No More Secondhand God
World Design Decade Documents
Operating Manual for Spaceship Earth
Utopia or Oblivion
The Buckminster Fuller Reader
I Seem to be a Verb
Buckminster Fuller to Children of Earth
Earth, Inc.
Synergetics: Explorations in the Geometry of Thinking
Synergetics 2: Further Explorations in the Geometry of Thinking
— both in collaboration with E. J. Applewhite
And It Came To Pass — Not To Stay
R. Buckminster Fuller on Education
Critical Path
Tetrascroll
The Grunch of Giants

INTUITION

by

R. BUCKMINSTER FULLER

Foreword by Norman Cousins

Second Edition

1983

Impact 〰 Publishers
POST OFFICE BOX 1094
SAN LUIS OBISPO, CALIFORNIA 93406

©Copyright 1970, 1972, 1973, 1983 by R. Buckminster Fuller

All rights reserved under International and Pan American Copyright Conventions. No part of this book may be reproduced, stored in a retrieval system, or transmitted in any form or by any means, electronic, mechanical, photocopying, recording or otherwise, without express written permission of the author or publisher, except for brief quotations in critical reviews.

INTUITION was first published by Doubleday & Co., Inc. in 1972; a revised edition was published by Anchor Press in 1973. The essay ''Mistake Mystique'' was first published in EAST/WEST JOURNAL in April, 1977, and reprinted in the book R. BUCKMINSTER FULLER ON EDUCATION, by the University of Massachusetts Press, 1979. It is reproduced here by the permission of the author.

The photographs of R. B. Fuller on the front cover and on the back cover at the helm of his sloop *Intuition* were taken by Martin Andrews, and are reprinted with his permission. The photograph of the Earth from space is courtesy of the National Aeronautics and Space Administration. The photograph of the sloop *Intuition* under sail is courtesy of the R. Buckminster Fuller Archives. The January 10, 1964, cover of TIME Magazine has been reprinted by permission from TIME. (©Copyright 1964 by TIME Inc. All rights reserved.)

This book was produced in collaboration with Antelope Island Press, P.O. Box 220, St. George, Utah 84770, whose invaluable contributions to acquisition, editing, design and production are gratefully acknowledged. Cover by Hal Hershey Book Production, Berkeley, California, based on a design by Robert E. Alberti. Printed in the United States of America.

Library of Congress Cataloging in Publication Data
Fuller, R. Buckminster (Richard Buckminster), 1895-
    Intuition.

    I. Title.
PS3511.U661715   1983   811'.54   83-8397
ISBN 0-915166-20-8 (pbk.)

Impact ☙ Publishers
POST OFFICE BOX 1094
SAN LUIS OBISPO, CALIFORNIA 93406

CONTENTS

Foreword

Bucky Fuller's hold on human beings is the product of many things. He is felt, and not merely heard. He is loved — not just because he is lovable but because he increases our pride in being human. He also creates new energies in us because he gives us an enlarged understanding of our relationship both to the universe and to things too small to see. His uniqueness as a teacher in these respects is that he sees poetry in everything and can communicate poetically. He views physics, astronomy, chemistry and other sciences as much through the creative imagination as through equations and formulas. In so doing, he refutes C. P. Snow's notion of the gulf between the ''two cultures''; he regards science and the fine arts as extensions of each other, as manifestations of an integrated reality.

If we read Bucky Fuller solely for information we will obtain information, but we will be cheating ourselves. We should read him for the enlarged awareness he gives us of our relationships to the universe around us; for the increased respect he gives us of human potentiality; for helping us to learn that there are no boundaries to learning and that infinity applies both to the cosmos and the human mind, which he celebrates above all else. We should also read him for the key he offers us, a key that can help unlock the innate wisdom — or *intuition* — that is such a rich source of natural wealth. In so doing, he helps us to mobilize our resources. There is a tendency in many of us to retreat from our own potentialities. Intuition is waiting to be put to use. Bucky gives us reason to trust it.

The great poets have attempted to describe the human mind and spirit but I doubt that any of them have been able to do so more provocatively than Bucky. The reason perhaps is that Bucky is not only inspired and nourished by the weightless and all-embracing entity called the human mind, but he has a way of opening our mind to the phenomena that lie within it. In this way, he introduces us to ourselves.

By opening the sluices of the human imagination to the universe around us, he helps us to do a better job of making our planet safer and more supportive than we have so far permitted it to be.

Norman Cousins
March, 1983

Introduction

When I was very young on our island in Maine, my father and mother gave me a sailboat when I could first swim. It was a very nice little sailing dinghy. One day my father and my mother and a girl cousin took my boat and went off from the island and sailed out of sight. They were going to another island several miles away to get our island's mail. All of a sudden I became terrifiedly aware that something had happened to them. It was true, a storm had come up and it was a bad day. With my experience sailing the boat, I had become sensitive to the fact that these were pretty bad conditions to cope with. It was raining when they took off, but they had wanted to get the very important mail, and so they had set out, wearing their storm oilskins. I ran from our island's harbor to the older people of the family at the other end of the island. I was then nine and was frighteningly certain that my father, mother and cousin were in trouble. Because I was so young, the rest of the family were not impressed. One of my cousins had married a minister, and he finally decided that they had best listen to Bucky. He decided to do something about it. He went down with me to our half-mile-away island harbor and got our captain, a native fisherman-farmer. We sent him off with our big 35-foot auxiliary engined Friendship sloop. Sure enough, he found my parents and cousin had been capsized and luckily had been able to swim to a nearby island. They were perishing with cold. The captain brought them home with my dinghy in tow.

That experience in my childhood made me absolutely certain of my clairvoyantly valid intuition that they were in peril. I think it happens to many, many children, perhaps built on all kinds of other experiences, such as when you are in the womb and things are happening to your mother outside there, and you are aware of it. What kinds of awareness do we have of things which transpire while we are in the womb? Are you aware of something outside? Probably, but not in geometrically conceptual terms since you have never yet seen anything.

I keep speculating on what, if anything, subconsciousness does while we are as yet in the womb. A newborn baby whose eyes are not yet open moves its fingers deftly. If you put your finger in its hand, the fingers close firmly on your finger. If you decide to move your finger away, the newborn baby's fingers immediately open. There is a real communication of touch already established at birth. I wonder what happens to unwanted babies while they are in the womb?

I myself have always held that my *intuition* is the most important faculty that I have.

This uniquely human quality has not always been highly regarded, however. In the 1920's world of science and technology, *intuition* became almost a "dirty word." In the years following their 1913 revolution, Soviet political leaders found it necessary to exercise absolute control over the economy in order to achieve full industrialization. They could allow no authority above that of the Communist Party, and thus banished religion and other metaphysical ideas — including intuition.

The atheistic position spread among the Western world's intellectuals as well, many of whom were assuming that the 1929 Crash and its abysmal recession phase in 1930 meant that capitalism was dead. Einstein's "Cosmic Religious Sense," published in the *New York Times Magazine* in 1930, was an inherently intuitive response to these trends.

In the U.S.S.R.'s 1920s idealogical war strategy, the word *intuition* was too dangerously mystical to be tolerated by an atheistic pragmatism. For a protracted period during the Great Depression, there was a general, Western world, intellectual, university-supported forsaking of accreditation of the concept and word *intuition* and other metaphysical-phenomena words. To the new communist pragmatists in the U.S.A., science did not use intuition; it simply stated its problems and solved them by empirical means.

I was very pleased when in the 1950s F.S.C. Northrop at Yale and a professor at another university, quite unbeknownst to one another, searched the personal papers of three great scientists who all the world agreed had made great physical knowledge discoveries on behalf of all humanity. Northrop and his independent counterpart examined the personal letters, personal scientific notes, and letters from the scientists' immediate family and friends, written only just before and slightly after the great scientists made their great discoveries.

Both research professors found, common to the writings of all these three great scientists who existed far apart in years and geography, that the single most important factor in their great breakthroughs was their *intuition*. The sudden strange impulse to look in this or that other direction, when physically busy over *here*, led to their great discoveries.

The second most important factor was a second intuition that occurred soon after they had made their great discoveries. The second intuition was always the awareness of what needed to be done "right now" if they wished to safely secure the great discovery they had just made as a practically realizable physical advantage for humanity. The second intuition might tell them, for instance, not to smoke another cigarette, not to go out to lunch, but to sit down right now and produce a clear documentation of the discovery. These first and second intuitions are analogous to those of a fisherman, alerted by the nibbling on baited hook and sinker, who realizes that he must jerk his line to secure the hooking and then must pull the line in and land the fish. All this first and second intuitioning was clear in the notes of these great discoverers. This published and widely read information of Professor Northrop powerfully reaccredited the intuition function amongst the scientific intellectuals.

To proliferate this knowledge I wrote my book *Intuition*. In it I find intuition and aesthetics to be two vitally important faculties of human's brains and minds.

Humans usually learn of the word *aesthetics* after their childhood. It is, however, pure aesthetics that makes a child love the sight of this or that. I think that I had personally very great good fortune in that my eyes at birth were and still are deformed to significant degree. I do not know what went on when I was in my mother's womb. At any rate, my eyes are misshapen. All that I have to do today is to take off my glasses if I wish to see what I saw until I was four-and-a-half years old. I have the same basic corrective lenses of my first 1899 prescription. Magnification has been added as I grow older. Without my glasses I cannot see anything as detailed as human eyes. For my first four-and-a-half years, I did not have glasses. All I could see were colors, brightness, shadows, darkness and outlines of large objects. As I take off my glasses now, all I can see is color. I was extremely sensitive to color. I was extremely sensitive to smell. I was extremely sensitive to touch. Not having the power of the eyes, which is so much greater than our other three faculties, imagine my astonishment, when they gave me glasses, to suddenly see eyes! I became so excited. I found that frogs had eyes, and toads had eyes, and snakes had eyes, and I kept looking deeply and intently into their eyes. They did the same to me. We seemed to say to each other, "I love and trust you."

So I went through a "second take" of life. It was like a rebirth. It gave me two kinds of ways of looking at things and therefore two different ways of thinking about my experience: in the hazy color way and in the detailed way.

I kept looking at human eyes and also snake's eyes and I found that the snake and I could talk to each other. As a child I would say intuitively that I have got to take this snake home because I love him. When my family or nurse began to take off my clothes, they would shriek because my clothes were full of snakes. I became very sensitive to what other people felt about different beautiful things. So, aesthetics to me was the live functioning of a flower, the ephemeral existence of a drop of water, and of the dew on the lilies. Fortunately, my childhood home had a lovely garden.

In exploring *intuition* and *aesthetics*, I've discovered something about how each one works.

Number one, we all have the experience of discovering and saying, ''What is the name of that mutual friend of ours? I can't quite remember it.'' Then the next morning I'll remember it. You've had that same experience. So we all have the experience of asking ourselves a question that gets answered by ourselves later on. In other words, this tells us we have a reliable subconscious functioning operative in our brains even while we sleep. Let's try to identify it.

Nowadays I'm traveling 90 percent of the time. With my very poor hearing sometimes I don't have the ability to hear the telephone or the alarm clock ringing. But, I am also often very tired; I swiftly reckon that I could sleep for perhaps three-and-a-quarter hours and then be awakened by that same subconscious faculty activity that retrieved the forgotten name for me while I slept. I intuitively dare do that if there is nobody, nor an adequate mechanical alarm, to wake me up. So what I do is to think up experiences I've had that are about three-hours-and-a-quarter long. Then I lie down and I wake up exactly three-hours-and-a-quarter later, right to the minute. This unconscious organization that goes on is employable by us willfully if we do some thinking about various magnitudes of time. I'm astonished to find I can wake myself in exactly three hours and 57 minutes if I want to. I've got to do a little thinking first to be sure I remember an experience that was exactly three hours and 57 minutes long, but it will work right to the minute. So what I have established now is that we have a subconscious monitoring process going on in our brain that employs experiences of various magnitudes, which can be reemployed.

Once I have established that we have a subconscious monitoring system, I then make another experiment. This has

13

to do with our "twilight zone" consciousness. I stretch both my arms forward and parallel to one another. I turn my two index fingers skyward. I now swing my outstretched arms horizontally away from one another with index fingers as yet skyward stretched. I keep looking straight ahead but realize that while looking straight ahead I can also see both of my skyward-pointed index fingers. I can see them clearly even when my arms are stretched out sideways, though I say to myself that I am looking straight ahead.

This experiment tells us that there are many events taking place in our lives at which we are not consciously looking but of which we are semiconsciously aware. It is this semiconsciousness which at times arouses our intuition to look directly in this and that direction and thereby discover some sensorially apprehensible phenomenon of significant relevance to our system's thinking. Semiconscious awareness of our experiences often constitutes a fourth event: the significant system-embracing event which produces comprehension of all the system's interrelationships, and even its relationships to other significant systems. Semiconscious awareness is a very effective and available basis for becoming aware of what is going on, in looking this way and looking that way.

It always takes a minimum of four equi-intervalled, identical, repeat-experiences to discover time dimensioning:

5 P.M.
Tuesday
January 6   Any one of many five
        o'clock experiences:
        having tea at X; met
        several people; many
        rememberable
        environmental items.

5 P.M.
Tuesday
February 3   "It seems to me that this
                 happened to me
                 before." "Oh, now I
                 remember, it must be
                 coincidence."

5 P.M.
Tuesday
March 3   "What again?" "How
                 long ago was that? And
                 the time before?" "If
                 this is not strange
                 coincidence, it will
                 happen again on March
                 31st."

5 P.M.
Tuesday
March 31   "Sure enough, I
                 have discovered a
                 periodic
                 experience."

Intuition relates to recognizing the significance of the "third" experience: That this is the third experience in which I felt intuitively that I've felt it before. The difference between the Leonardo da Vincis and Albert Einsteins and other people is that the other people don't pay attention to the third experience in the series of four system-defining events. The Einsteins do.

We now know that all human beings have an electromagnetic field of incredibly high frequency and low voltage. It is very difficult to read, but it is there. And when we are feeling negative, it produces a negative field, and when we are feeling positive, it produces a positive field. This may be why people drink alcohol together — to release the positive "good fellowship field" among them. Thus far, we don't know very much about these ultra, ultra shortwave phenomena.

My main purpose here is to give you concepts about aesthetics and intuition. They are probably all interrelated.

In the matter of aesthetics consider our sense of smell. The smell of healthy growth — a new baby, fresh biscuits — gives a smelling sense of growth vs. decay. This also has something to do with our sense of balance and our spontaneous gauge of accuracy in design. Our sense of convergence and divergence must augment our appraisal of growth and decay. I think that the great sculptors of the era of the great sculpture in Greece and Italy had an incredible aesthetic sense of structure and function, of functional adequacy and of proportion as manifest, for example, in drawing the neck or back of a woman. I see that this judgment of good design and accuracy often comes while looking at flowers, petals and stems. Our sense of shape or of compound curvature is keen.

A structural column when top-loaded tends to bend and twist in its mid-height. Structural engineering measures and employs the "slenderness" ratio of a column's mid-height cross-section diameter as ratioed to the column's overall height. The Greeks' temple columns were made of stone cylinders stacked vertically. The limit slenderness ratio of a Greek column was eighteen-to-one, i.e., eighteen column diameters high. Present-day steel columns have a slenderness ratio of 40 diameters high. What I identify as "aesthetic discernment" in human males is their ability to appraise the relative perfection of the slenderness ratio in a female's leg and to judge the variation of diameters to within a 32nd of an inch as seen from the 25-yard distance. The sense of verticality error of picture hanging and of ship or flagpole masts is also accurate to an unbelievably meager unit of measurement error.

In biological structuring, the compressional function is accomplished by hydraulics. The crystalline intertensioning contains the compressional hydraulic stresses. Liquids distribute their loads. Crystals will not distribute a load but liquids do. Gases are compressible. Liquids are flexible but noncompressible. Liquids, then, guarantee the firmness of all biological, botanical and zoological structuring.

All human beings are over 60 percent water. We have crystalline tensional sacks filled with liquid. The sacks overlap each other like fibers spun into thread. The liquid-filled sacks are often held together by spiraling-crystal intertensioning. We have a great sense of how much liquids are coming in or out of our system. We have structural understanding propensities.

We *feel* the structural integrity of triangles. If we make a large necklace with four or more (twenty or any other large number) of six-inch long aluminum tubes strung together on a dacron line, we will find it very flexible, which means structurally unstable. We find that the flexibility does not involve bending the tubes or changing the lengths of the tubes. We find that the flexing is done by the short lengths of dacron cord running between the tubes' ends. Hoping to isolate the flexing phenomenon, we keep taking tubes out of our necklace. When we get to only five, or even four tubes being left in the necklace, it is as yet completely flexible. We now oust one more tube and that is the last that we can take out and have the assembly produce a necklace having an opening through it. The Greeks used the name ''polygon'' for planar-pattern geometrical figures. Six, five or four do not hold any regular shape. They fold together. There is no such thing as a square. The three final tubes left in the configuration produce a triangle. It is the first and only polygon to hold its shape. It is also the minimum, terminal case of a polygon series.

The triangle, which holds its shape despite flexible corners, is structure. I define structure as a complex of events that interact to produce a stable pattern. I define a system as that which has an insideness and an outsideness. There are only three structural, omni-equi-angle triangle-enclosed systems in Universe — the four-vertex, six-vertex, and twelve-vertex systems. In old Greek, they are called the tetrahedron, octahedron, and icosahedron. Altogether, all crystals and all biological design are made up from those three. When we fasten three corners of a triangular face of any one of the three prime structural systems of Universe together, they form a rigid crystal. When we bring two corners of each of the prime structural systems together, they form a hinge and produce the behavior of a liquid that can articulate and distribute stresses but never compress. If we bring one corner of each of two of our prime structural systems together, they will bind to one another in an open stress-distributing but compressible system as a gas.

I think aesthetics is built on an innate sense of the stress distribution, flexing, and rigidifying capabilities. I am the first to show why these system interbondings are single-bondedly the gases, double-bondedly the liquids, and triple-bondedly the crystals. We have a Universe outside (macro) and a Universe inside (micro) set of systems interbonded to produce all physical matter. Our thoughts, our experiences, the book, the table — all are systems.

I am convinced that every child is born a genius. Most are degeniused by loving parents who are afraid the genius-inspired initiative of their children may get them in trouble with the going socioeconomic system in which they live.

Life begins with awareness, and with no otherness there is no awareness. Life begins when you get outside — come out of the insideness and of the womb and become both insideness and outsideness and self-observing — and realize the outsideness and insideness of the otherness. Our innate genius led me to discovering a system and the structural mathematics of systems. The phenomenon which I call *synergetics* is now beginning to be recognized and is upsetting all academic mathematics and science. I have enough confidence in my experiences to date to rely on my natural *intuition* and my natural aesthetic feelings which led me to making experiments and thereby acquiring experimental evidence of the structuring theretofore only intuitively dreamed to be possible.

Intuition often turns dreams into demonstrable facts.

R. Buckminster Fuller
March, 1983

INTUITION—

Metaphysical Mosaic

Intuition —

Metaphysical Mosaic

The following thoughts regarding
the acquisition, commissioning and naming
of a new seventeen-ton
ocean-cruising sloop
occurred and were inscribed
throughout the morning hours,
immediately preceding the moment
at which the craft was lowered
by a giant motorized sling
into her design destined
REALIZATION
of waterborne existence
at midday July 31, 1968.

Life's original event
and the game of life's
order of play
are involuntarily initiated,
and inherently subject to modification
by the a priori mystery,
within which consciousness first formulates
and from which enveloping and permeating mystery
consciousness never completely separates,

but which it often ignores
then forgets altogether
or deliberately disdains.
And consciousness begins
as an awareness of otherness,
which otherness-awareness requires time.
And all statements by consciousness
are in the comparative terms
of prior observations of consciousness
("it's warmer, it's quicker, it's bigger
than the other or others").
Minimal consciousness evokes time,
as a nonsimultaneous sequence of experiences.
Consciousness dawns
with the second experience.
This is why consciousness
identified the basic increment of time
as being a *second*.

Not until the second experience
did time and consciousness
combine as human life.

Time, relativity and consciousness
are always and only coexistent functions
of an a priori Universe,
which, beginning with the twoness of secondness,
is inherently plural.

Ergo:
All monological explanations of Universe
are inherently inadequate
and axiomatically fallacious.

There can be no *single key*
nor *unit building block* of Universe.

Before humans had learned by experiment
that light and other radiation
have unique speed,
humans thought of sight
as being instantaneous.
Under this misapprehension
classical science assumed
that every action had a *simultaneous* reaction.
Since light *does* have a speed,
and there is no simultaneity,
there are inherent event lags.
We learn that every event is triplex,
consisting of an *action*
that has both a *reaction* and a *resultant*.
These three inseparable functions of an event
are nonidentical
and nonsimultaneous.
In the event called ''diesel ship-at-sea''
the action of the ship's propeller
has a thrust pattern
to which the ship reacts by moving forward,
which also results, secondarily,
in the ship's bow elevated wave,
and its depressed transverse stern wave
which wave disturbances of the water
are separate from the propeller's thrust wave.

Ships appear to be so solid
that they negate human perception
of their minute longitudinal contraction,

which occurs initially as a consequence
of the interaction of the ship's inertia
with the propeller's thrust.
This contraction
and its subsequent expansion
could be observed in yesterday's
loosely coupled railway trains,
as they jerkingly accelerated or stopped.

In addition to inherent duality of Universe
there is also and always
an inherent *threefoldedness* and *fourfoldedness*
of initial consciousness
and of all experience.
For in addition to (1) action, (2) reaction, (3) resultant,
there is always (4) the a priori environment,
within which the event occurs,
i.e., the *at-first-nothingness* around us
of the child graduated from the womb,
within which seeming nothingness (fourthness)
the inherently threefold
local *event* took place.

Whether our experience episodes
are voluntary or involuntary,
passive or active,
subjective or objective,
our brains always and only
isolate, tune-in,
modulate and document,
store, retrieve and compare *informedly*,
or speculatively formulate,
in special-case increments of unique concepts.

God gave humans a faculty
beyond that of their and other creatures'
magnificent physical brains —
and that unique faculty
is the metaphysically operative mind.

Brains apprehend and register
store and retrieve
the sensorial information
regarding each special-case experience.

Mind alone can and does
discover heretofore unknown
integral pattern concepts
and generalized principles,
apparently holding true
throughout whole fields of experience.
And once discovered by mind
the concepts of the generalized principles
become additional special-case experiences,
and are stored in the brain bank
and are retrievable thereafter by the brain.
But brains and their externalized
detachedly operating descendants —
the electronic computers —
can only search out and program
the *already experienced* concepts,
and mind alone can recognize and capture
the unknown and unexpectedly existent,
ergo, unsearchable, unwatched-for
generalized principles.
If you do not know
the behaviors exist,

you cannot be
on watch for them.

Weightless, perceptive, prescient mind
alone enabled humanity
also to conceive of new, original
and objective ways to employ
the (only subjectively acquired) concepts
of generalized principles,
such for instance as *leverage*,
which empowered men
to conceive of practical ways
to both elevate and move
objects manyfold their own weights,
or that of their direct muscles'
lifting, pushing and pulling abilities.

And this greatly augmented
humanity's competence
to heed anticipatorily
the lessons of past negative experiences,
and with enlightened logic
to alter the environment
in ways permitted by nature,
which would protect humanity
against external and internal deprivations
while also increasing the sustenance
of increasing numbers of humans
for increasing numbers of days
of their potential life spans.

Or mind enables
a co-operative succession of humans
both to discover and objectively employ

a complex family
of generalized principles
brought from
the weightless, timeless,
metaphysical integrity and fidelity
of absolutely orderly
eternal Universe;
brought into
time and energy synchronized consciousness
of the physical evolution scenario;
brought by
a plurality of individually
and remotely operating —
but interregeneratively —
inspiring and educating
exquisitely prescient minds.

A typical family
of generalized laws,
thus interinspiringly discovered
by a plurality of human minds,
would start with that first disclosed by Avogadro
of a constant number of molecules
in a given volume
of any and all chemical elements
isolatable in their gaseous state
under identical conditions
of heat and pressure,
which mathematically
statable law
compounded with Boyle's chemical law
to informedly inspire Mendéleyev
and a few other colleagues
to differentiate out and predict

the existence of a closed family
of ninety-two
regenerative chemical elements.
These elements, when found,
they said would display
such-and-such
unique and orderly characteristics,
and they mathematically identified in advance
the respective constituent quantities
of the as yet undiscovered discrete characteristics
of the as yet undiscovered elements.
All the members of this interregenerative
information relay team
did not know one another personally.
Their accumulatingly inspired prediction occurred
at the historical moment
when, a century ago —
only fifty-two
of the ninety-two
of those heretofore unexpected chemical elements
had as yet been discovered
and physically isolated
by humans on Earth.

Sir James Jeans said that
science is
the attempt to set in order
the facts of experience.
Mendéleyev, attempting scientifically
to find an order
in which to set
the first fifty-two chemical elements,
inadvertently uncovered
a previously unknown

system of regularities
common to all fifty-two,
which, if their implied generalization
proved in due course
to hold true,
would require the presence in Universe
of the additional unknown forty
to fill in the membership vacancies
occurring in the revealed periodic behaviors
of the already discovered chemical elements.

Since Mendéleyev's prediction,
every few years —
one by one —
all ninety-two have been identified
as being present in various abundances
in all the known stars of the heavens,
while ninety-one of them
have been isolated by scientists
somewhere on planet Earth,
and all of them have the exact characteristics
predicted by Mendéleyev and his colleagues.

And all the foregoing
subjective harvesting —
accomplished by individuals
bound together
by naught other than intellectual integrity —
has enabled still other
remotely exploring,
educatively inspired individuals
first to discover,
then inventively to employ
the originally unknown

uniquely recombining
synergetic behaviors —
in structural groupings —
of those ninety-two regenerative elements,
thereby attaining utterly surprising
structural, mechanical,
chemical and electromagnetic characteristics
which have enormously increased
the relative advantage
of ever-increasing numbers of humans
to cope with the challenges of life
by accomplishing ever more difficult tasks,
previously considered impossible to do;
with ever less
time, weight and energy investments,
augmented exponentially
by ever-greater investment of unweighables
of the metaphysical resources —
of hours of thoughtful reconsiderations,
anticipations, conceptualizing,
searchings and researchings,
calculations and experiments.

But despite mind's ability
to capture, mathematically equate
and employ in conscious temporality
the eternal weightless generalized principles,
Humans cannot design
a *generalized* anything.
We can only embody
the eternal design's generalized principles
in *special-case* employments
of those *generalized principles*.

In confirmation of the foregoing
we note that
while Archimedes discovered
the generalized principle
governing *displacement*
of all floating bodies —
in respect to the flotation medium
in terms of their respective volume-weight ratios,
we cannot design
a generalized boat.
It must be a specific canoe,
a ferryboat, or sloop —
and each of unique size and capability and durability
for all special-case embodiments
are entropically fated
to disintegrate in time.
Whether the experience episodes
are passive or active —
i.e., involuntary or voluntary,
subjective or objective —
brain always and only
isolates, tunes in, documents,
stores and retrieves
special-case concepts.
Only mind can discover,
comprehend, equate and employ
the absolutely weightless,
ergo, purely metaphysical,
generalized principles
which, being weightless and unfailing,
must be eternal.

Synergy is one
of those generalized principles.

It is defined scientifically
as behavior of whole systems
unpredicted by behaviors
of any of their separate parts.
Synergy is disclosed, for instance,
by the attraction for one another
of two or more separate objects.

Such objects, however,
on closer inspection
are themselves mass-attractively integrated
energy event aggregates,
each of which is so closely amassed
as to be superficially deceptive
and therefore misidentified
by humanity's optically limited discernment,
as *bodies* —
separate "solid" bodies —
despite that physics has never found
any "solid" phenomena.

For instances of unpredictable
synergetically mass-interattracted
and mutually interco-ordinating phenomena,
we witness the mathematical regularity
with which objects
accelerate toward Earth —
or, as it is said,
"fall" into Earth.
Or we witness the co-ordinate behaviors
of the corotating, co-orbiting
Earth and its Moon,
or of any two or more neighboring,

larger or smaller,
uniquely amassed,
separatedly and individually existing,
harmonically frequenced,
neighboring aggregates
being both mass attractively
and precessionally reassociative
in the manner superficially identified by humans
as *matter* —
be they metallic or nonmetallic.

Another intuitively interinspired
experimentally informed
scientific relay team's
century-spanning accumulative accomplishment
by remotely existing individuals,
started with Tycho Brahe's
instrumentally harvested data
on the Sun's planets.

The *mass attraction* of those remotely separated,
seemingly *solid* planets —
which we now know to be
microcosmic aggregates
of energy events —
was first hypothetically explained
by Kepler
to account for the geometrical regularities
of interco-ordination
of their elliptical orbits
by the Sun's planets.
Kepler found the regularity
for which he searched

in the identical areas
of the different
pie-shaped segments of the sky —
some short and wide,
some long and thin —
"swept-out" in a given time
by imaginary radial tethers
tied to each of the planets
from the same star-Sun center
around which they traveled,
each at vastly different distances
and at vastly different rates.

"How and why," asked Kepler,
"could these separate planets
act in such unitary co-ordination
with one another and the Sun,
without any visible
mechanical or structural
interconnectors?"

Inspired by Brahe and Kepler,
as well as by Galileo's
rate-of-fall measurements
and law-of-motion formulations,
Isaac Newton hypothesized
that a body should move
in an astronomically straight line
except as affected
by other celestial bodies,
such, for instance, as manifest
by Kepler's hypothetical
interplanetary attractiveness.

Newton calculated the straight line
tangential to its orbiting,
along which the Moon would move
if it were not attracted by the Earth,
and measured the rate at which the Moon
fell away from that line
in toward the Earth.
He found that the rate of falling
corresponded exactly
with Galileo's observation
of an accelerating-acceleration
in the rate of Earthward fall,
which elegance of scientific agreement
reinspired Newton
to evolve his mass attraction law,
which showed that the relative initial value
of Kepler's assumed mass attraction
could be determined
by multiplying the two interattracted masses
by one another
and increasing the attraction value fourfold
each time the distance
between the two bodies is halved.
This explained Galileo's observed
accelerating-acceleration
of bodies falling inward
toward Earth.

Tested by the interim centuries,
Newton's law has since come to explain
the interattraction integrities
of all macro and micro behaviors of Universe.

Because science has not found any property
of any one of the bodies
which, considered only by itself,
predicts that the body will attract,
or be attracted by another body —
and, even more surprisingly,
with the rate of that attraction
increasing exponentially
as they approach one another —
it is in experimentally demonstrated evidence
that neither the phenomenon mass attraction,
nor its even more surprising
second power, algebraic rate
of interattraction increase,
can be disclosed
and humanly comprehended
only by observation
of the integral body characteristics
of any one body
while disregarding
the mutually covarying behaviors
of both bodies,
or of all of a complex of bodies
comprising the observed system,
as comprehensively and progressively measured
and mathematically described
in terms of relative proximity,
relative mass and relative dimensionings,
relative velocities
and their respective
rates of change.

Since both mass attraction
and accelerating-acceleration

are experimentally demonstrable,
while no property of any one part
has been discovered
which predicts either the attraction
or its accelerative gain —
and in fact
no property of one part
considered only by itself
predicts the existence of another part —
synergy is experimentally —
which means scientifically —
manifest.
Q.E.D.

Synergy is the *only* word that means
behaviors of a whole system
unpredicted by the separate behaviors
of any of its parts.

As of 1970,
world-around questioning
of three hundred university audiences,
averaging five hundred persons each,
finds the word synergy —
as well as the phenomenon it identifies —
to be known to less than three percent,
of university students or staffs.

As of 1970 also,
persistent inquiry of general public audiences
discloses a knowledge of synergy
by only one percent of general society.

Because all the generalized principles

thus far discovered
are uniquely identifiable
exclusively as
"behavioral interrelationships"
of *two or more*
separate components —
which, in turn, consist
of energy event-aggregates —
the generalized principles
are not to be confused
with the profusion of data
concerning common characteristics or statistics
of separate components of systems,
which are myopically
and oversimplifyingly observed
by an everywhere specializing society.

Such common statistics
can be monitored satisfactorily
by the brain or
its externalized adjunct,
the computer;
but unknown synergies
of as yet unknown generalized principles
which are inherently and exclusively,
integral behaviors of two or more,
cannot be programmed for computer discovery.
Because the relevant unknown behaviors
are experimentally demonstrable only in retrospect
as existing only *between*
but not *of* or *in* any one part,
the synergetic behaviors
of a plurality of parts
are inherently unpredictable.

Development is programable;
discovery is not programable.
Since the behaviors to be sought
are unknown,
computers cannot be instructed
to watch out for them.
Computers can ''keep track''
of a complex of behaviors,
but only human mind can discern
the heretofore unknown
unique interrelationships
which exist *between* and not *of*
the separate bodies.

Even less known and understood
than the generalized principle
mass attraction
is the generalized principle of *precession*
and the synergetic phenomenon
which it uniquely identifies.

Precession is the behavioral interrelationship
of remote and differently velocitied,
differently directioned,
and independently moving bodies
upon one another's separate motions
and motion interpatternings.

Mass attraction is to *precession*
as a *single note* is to *music*.
Precession is *angularly accelerating*,
regeneratively progressive
mass attraction.

Because the Sun's planets
did not fall into one another
Kepler's discovery of their elliptic orbiting
as well as the solar system's motion
relative to other star groups
of the Galactic Nebula
are all and only accounted for
by *precession*.

Precession is uniquely dependent
upon the entirely unexplained,
ergo mystically occurring,
omnimotions of Universe
successfully hypothesized by Einstein
in contradistinction
to Newton's assumed
a priori cosmic norm of ''at rest.''

Precession is a *second-degree synergy*
because it is not predicted by mass attraction
considered only by itself.
Mass attraction
is experienced intimately by Earthians
as gravity's pulling
inward toward Earth's center
any and all objects
within critical proximity
to Earth's surface,
and moving through space
at approximately the same speed
and in the same direction
as those of planet Earth.

Not until we learn by observation

that the mass attraction
of any two, noncritically proximate
bodies in motion
imposes a motional direction
at ninety degrees
to their interattraction axis,
do we learn of this second surprise behavior
of two or more bodies.
They no longer ''fall-in,''
one to the other.

Thus is the Moon
precessed into elliptical orbit about the Earth
as the Earth and Moon, together,
are precessed into elliptical orbit around the Sun,
yielding only in a ninety-degree direction
to the Sun's massive pull —
being beyond
the critical proximity distances
for falling into one another.

Unlike ninety-nine point nine nine nine
percent of all humans,
Goddard, carefully heeding the laws
of both mass attraction and precession,
realized that an object,
rocket-propelled or accelerated,
into a different velocity —
and into a different direction
to that of the Earth's
speed and course around the Sun —
would have its gravitational pull
toward the Earth
reduced fourfold

every time it doubled
its distance away from the Earth;
only a hundred miles out
from our Earth's surface
the attraction would be
so diminished
that it would permit the Moon's pull
to become significant,
at which distance
the rocketed object
would lose its tendency
to fall back into the Earth,
and now affected dominantly
by the integrated mass attractions
of all other celestial bodies,
would go into orbit
around our Earth.

And to understand
how little is that
one-hundred-mile distance
out from Earth's surface
at which *orbiting*
replaces the tendency
to *fall* back into the Earth,
we note that
the thickness of a matchstick
out from the surface
of a twelve-inch diametered
household "World Globe"
is the distance at which
our first rocketed objects
do go into orbit.

Mass attraction and precession
provide the first scientific means
of elucidating social behavior.
When humans affect one another
metaphysically,
the least thoughtful
goes into local system orbit
around the most thoughtful.
When humans tense one another
physically,
the least strong
falls into the other,
"falls" in love.
When they repel one another physically
the least strong is rocketed
into remote system orbit.

Because the physical characteristics
of an aggregate's separate components
and their respective submotions
cannot explain the behaviors
of their progressively encompassing
and progressively complex systems,
we learn that
there are progressive degrees of synergy,
that is to say,
synergy-of-synergies,
which means
complexes of behavior aggregates
holistically unpredicted
by the separate behaviors
of any of their subcomplex-aggregates;
and because mass attraction

does not predict precession
each subcomplex-aggregate
is in itself
only a component behavioral aggregation
within an even greater
behavioral aggregation,
whose comprehensive behaviors
are never predicted
by the component-aggregates alone.
It is, furthermore,
in experimentally disclosed evidence
that there is
a synergetic progression in Universe —
a hierarchy of total complex behaviors
entirely unpredicted
by their successive
subcomplexes' behaviors.

This means that there exists
a synergetic progression
of ever more encompassing systems
of human experience discernibility
which are spontaneously differentiated
into unique levels
of cognitory consideration
in which the contained micro
of any adjacent macro level
never predicts the existence
or the observed behaviors
of the adjacently next most encompassing
macro level complex.

Thus are the *atoms*
unpredicted by any

of their individual
neutrons, protons,
positrons, electrons,
neutrinos and antineutrinos
et al.

Nor does any one atom
in itself predict
the family of
periodically co-ordinate
unique chemical elements
in ninety-two, self-regenerative varieties
of mathematically incisive order;
together with their several hundred
of interspersed isotopes
to be coexistent
in complex but orderly array.

And the periodic behaviors
of the chemical elements
and their isotopes
in turn fail to predict
their aggregate behaviors
as molecular structurings
of various, harmonically complexed,
unique associabilities of atoms,
known as chemical compounds.
Nor do the chemical compounds' molecular structures
have inherent characteristics
which predict the level
of organic associability
of molecules as biological cells.
And the level of biological cells
does not predict

their association in turn
as biological tissue —
the first human
naked eye-discernible level —
of these synergetic behaviors.

And the level of tissues
does not predict
organic biological species
in a vast variety
of permitted design alternatives,
whose unique pattern structurings
are chromosomically programmed
to replacingly aggregate
as regenerative organisms
ecologically interacting
all around our planet
in chemical phase intercomplementations
all fundamentally actuated
by a combination of mathematical symmetries and cycles
all pyramided upon
mass interattraction of atoms
dynamically hovering in orbits
within critical "fall-in" proximity
of one another.
Or fallen into
a single-bonded *fluttering*
or into double-bonded *hinging*
or into triple-bonded *rigidity*
or into quadruple-bonded *densification* —
ergo all pyramided upon
exponentially compounded
synergies of *mass attraction* and *precession*.

When adequate acceleration
is imparted to micro aggregations
of atoms,
sufficient for them to escape
the critical limits
of both mass attraction
and precession intereffects,
then radiation
at 186,000 miles per second
of the separate energy quanta
ensues,
and the generalized behavioral law
is that cited by Einstein's
$E = Mc^2$.

Inasmuch as *mass attraction*
and its second-degree synergy, *precession,*
together with *radiation*, most prominently explain
the *synergetic* interpatterning integrity
of both macro and micro
aspects of Universe,
and the fact that
three out of four of the names
of their behavioral identities
are popularly unfamiliar,
provides experimental evidence
that less than one percent of humanity
has the slightest notion
regarding the extraordinary principles
kinetically structuring and cohering
the integrity of eternally regenerative Universe.

While one percent of society
has superficial awareness

of the existence of mathematical regularities
synergetically displayed by *mass attraction*
and supersynergetically displayed as *precession*,
no scientist has the slightest idea
what *mass attraction* is
nor why
synergy, precession or *radiation*
exist or act as they do.

Nobel laureate physicists,
in self-conscious defense
of their abruptly discovered ignorance
in regard to such cosmically important matters
(understanding knowledge of which
society has accredited them with possessing)
shrug off the necessity to explain
by saying, ''Here we will have to assume
some angels to be pushing things around.''
Though popularly unrealized,
it is in experimental evidence
that the origins of science
are inherently immersed
in an a priori mystery.

This explains why
the history of science
is a history of
unpredicted discoveries
and will continue
so to be.

But within the mystery
lies the region

of humanly discovered phenomena
whose whole region
is progressively disclosing
an omni-integrity of orderliness,
of interactive and interaccommodative
generalized principles.

In view of all the foregoing
the preponderance of as yet
scientifically uninformed peoples
explains why human awareness
has at first and for long
greatly misapprehended,
for instance, as "solids"
the superficially deceptive microaggregates
which defied differentiating resolution,
into their myriads of separate parts,
by the instrumentally unaided
human sight.

Thus loving humans
have unwittingly tutored
their young to acquire
a whole body of reflexes
labeled as knowledge,
all of which has since been invalidated
by experimental science's findings —
as armed with powerful instruments
for exploring
the ninety-nine percent of reality,
which is inherently
untunable directly by the human senses,
humans grope for *absolute* understanding,

unmindful of the a priori mystery
which inherently precludes
absolute understanding.
Unaware that their groping
does not signify personal deficiency,
and ignorant of the scientific disclosure
of fundamentally inherent mystery,
they try to ''cover up'' their ignorance
by asserting that no fundamental mystery exists.

The omnicommitment
of the twentieth century's
world-around society
to the synergy invalidated misconception
that specialization
is desirable and inevitable,
tends to preclude humanity's
swift realization
of its many misconceptionings
and its necessity to substitute therefore
tactically reliable information.
Specialization is antisynergy.

In short, physics has discovered
that there are no solids,
no continuous surfaces,
no straight lines;
only waves,
no things
only energy *event* complexes,
only behaviors,
only verbs,
only relationships,

which, once discovered,
can be kept track of
and employed
by both the integral brain
and the extracorporeal computer
but may never be discovered originally
by those physically limited tools.

And the *why-for* and *how-come*
of omni-interaccommodation
of all the known family
of weightless, eternal, generalized principles —
thus far discovered
by scientific observation
to be metaphysically governing
in elegant mathematical order
all Scenario Universe's
interrelationships, transformations and transactions,
without one principle contradicting another —
are all and together
absolute mystery.

Offsetting the formidable dilemma
of comprehensive social ignorance
human mind finds
a new comprehending advantage
to be inherent in the discovery
that within all the foregoing
progressively encompassing
hierarchy of synergetic levels,
each encompassing level does manifest
synergetic behaviors
unpredicted by the behaviors

of any of its sublevels'
components' behaviors —
considered only by themselves.
Though neither known nor anticipated
by the status quo's present body of knowledge
this hierarchy of hierarchies
constitutes a cosmic consistency,
warranting its recognition
as a generalized law of Universe.

Knowledge is of the brain,
wisdom is of the mind,
and there is herewith implicit
an a priori wisdom-of-wisdoms.

Out of the a priori mystery
from time to time
mind fishes a new
generalized principle,
which though absolutely unique
always accommodates and integrates
with all the previously discovered
generalized principles.

All of which are originally apprehensible
only by weightless mind
which alone of all phenomena
can cope knowingly
with the eternality of principles.

The omni-interaccommodativeness
of the totally known inventory
of generalized principles
constitutes progressive disclosure

of a vast a priori design
to be governing Universe,
whose intellectual integrity bespeaks
an a priori greater intellect
than that manifest in humans,
all of which synergetic integral
is hidden from sight of humanity,
as it is at present omnivictimized
by a universally specializing antisynergetic,
anticosmological
educational process.

Wherefore it is also manifest that
Universe is the maximum synergy-of-synergies
being utterly unpredicted by any of its parts
or by the hierarchy of synergies
of ever exponentially advancing degree —
no complex stage
having been predicted
by its parts.
For instance,
the chemistry and structure
of the human's toenail
in no way predicts
the complex, organic behavior
known synergistically
as humans.

For humanity to comprehend
in individually effective degree
the life in Universe
which it is experiencing,
it is logically required
that humanity must develop

self-initiated and self-disciplined reconsiderations
of its total inventory of experiences
from which it may, hopefully,
gain a degree of knowledge
adequate to humanity's
spontaneous initiation
of conscious and competent co-operation
with the orderly processes
of universal evolution.

In order to accommodate
and abet evolution,
humanity must heed
the synergistic hierarchy
and must commence by heeding
that human mind has discovered
and proved mathematically,
two millennia ago
in Ionian Greece,
a generalized principle —
corollary to synergy —
whose mathematical characteristics are statable as:
the known behavior
of a whole system
and the known behavior
of some of its parts
make possible the discovery
of other — if not all —
of the originally *unknown*
component parts
of the system.
Therefore in direct contradiction to present specialization,
all educational processes

must henceforth commence
at the most comprehensive level
of mental preoccupation,
and that level is the one
that consists of the earnest attempt
to embrace the whole eternally regenerative phenomenon
Scenario Universe.

And this is what children
try to do spontaneously
whenever they ask their parents
embarrassingly important
cosmological questions.
Evolution, most powerfully operative
in the metaphysical spontaneity of children,
is trying to break through
the barrier of ignorance of synergy and mystery
which as yet frustrate human understanding.
Perversely, the parents
tell them to forget Universe
and to concentrate
with A,B,C,
1,2 and 3 —
only with parts,
which process the parents
call "Elementary Education,"
and reflexively misconceive
as the essential beginning
of all learning processes.

In our now to be adopted
synergetically strategied
educational process

we are aided
by scientists who preceded us,
for Euler discovered
early in the nineteenth century,
that all patterns of Universe
can be resolved into three
conceptual differentiations:
lines, crossings and *areas.*
Systems divide Universe
into a plurality of regions:
all of the Universe
which is outside the system
and all of the Universe
which is embraced by the system.
Systems are inherently polyhedronal.
Systems of thought
divide the Universe
into the conceptual and nonconceptual.
Conceptual systems always consist
of a constant relative abundance
of the *lines, crossings* and *areas*
in which $C + A = L + 2$.

And because of this
constant relative abundance
whole pattern behaviors
of all our experiences —
when properly conceptioned —
can be comprehensively differentiated,
topologically equated, observed and considered.
Thus was Einstein synergetically
and equatingly advantaged, with cosmological integration
of the known behaviors of the whole physical Universe,

as well as of behaviors of its two principal parts,
mathematically to predict and discover
heretofore unknown
and unexpected behaviors
of heretofore unrecognized
constituents of the whole.

Intuitively stimulated
by experimentally demonstrable
speed of radiation knowledge,
as well as by Brownian movement
and black body heat,
Einstein started holistically
with the concept of Scenario Universe
as an aggregate
of nonsimultaneous,
complexedly frequencied,
and only partially overlapping
ever and everywhere
methodically intertransforming events
which conceptioning
is superbly illustrated by an evening
of overlappingly frequenced fireworks.
$E = Mc^2$
and
$C + A = L + 2$
experimentally conceptualized
by those partially overlapping
fireworks events
as one rocket is blasted off
before the previous rocket's
unique display has been completed,
and both a moving picture camera

and a tripodded still camera
can be set
with their lenses left open
to register the whole evening's
fireworks program
both as a scenario
and as single, composite, static pictures
of all the light patternings
that took place
against the black void of sky.

And the synergetic relationships
of the scenario footage
and the "still" photographs
together may become
the basic experimental evidence
of fundamental self-education.

For E = physical Universe,
which consists of energy
in two phases:
(*One*) *Energy associative,*
as matter = M,
where critical proximity
accounts for all the atoms
either falling into one another
or precessing into very local orbits.
(*Two*) *Energy disassociative,*
as radiation = c^2,
for the omnidirectional light-wave growth sphere
increases as the second power
of the linear speed of light.

c = linear speed of all radiation
c^2 = radiant growth rate of a special wave
The radiant light discloses the
trajectory lines of the successive
rocket blast-offs
whose trajectory lines = L
cross one another = C,
as the local "burst" lines
complexedly define areas = A
and the whole fireworks
demonstrate the patterning
of Einstein's Universe as a Scenario Universe —
of "nonsimultaneous and only partially overlapping
transformation-events."
Again Q.E.D.

And the black void
nothingness of night
backdropping the fireworks
is the omnipresent,
a priori mystery.
And the real beginning of education
must be the experimental realization
of absolute mystery.

For the a priori
comprehensive and permeative
mystery of Universe
is approximately unknown,
or is deliberately side-stepped,
or is just overlooked
by most educators,

and is politically acknowledged
only as orthodox religions.

And again
society's lack of knowledge
of the a priori mystery,
and its pragmatic conditioning of its reflexes
by leaving to its priests
what manner of response
they should make
to the innate intuitive awareness
of the a priori mystery,
permit the persistence
of such ignorant cerebrations
as that which for instance
invents atheism.
Tending to discount science
society has no working knowledge
which contradicts such assumption.
The scientific proof of synergy
is, however, experimentally demonstrable
and is experienced
in a myriad of ways.
As for instance
in the tensile strengths of alloyed metals,
such as chrome-nickel-steel —
which is severalfold stronger
than the sum of the tensile strengths
of each of the separate metals
which altogether comprise the alloy —
metallic alloying is explained only by synergy.

Such high heats and stresses
were involved in a jet engine

that it could not be realized
until chrome-nickel-steel
was discovered and produced.
And the jet engine
within only one decade of years
has shrunk our Earth
into a one-town dimension,
which now realized accomplishment
was mysteriously unanticipated
by any scientific society
of yesterday's
governments, corporations,
educators and politicians,
who even now utterly disregard
the mysterious realization
that synergy now permits
the logically predictable
humanly conceivable and executable
rearrangements of environmental constituents,
in ways which are sufficiently favorable
for the regeneration of all life
aboard our planet Earth
by producing ever more performance
with ever less pounds, minutes and watts
per each function served.

And the more-with-lessing
constitutes ever-increasing mastery
of physical behaviors of Universe
by the metaphysically operative verb
mind:
and all the foregoing
implies incontrovertibly
the progressive realization

by humans on Earth
not only of a vast
universal design
but of a Universe Scenario,
whose a priori conceptioning
is clearly intent
to render Earth-riding humans
a comprehensive physical success,
despite humanity's
as yet undiscarded
ignorance, fear
and distrust of its mind.

But humans are designed,
(again a priori)
metaphysically equipped, and advantaged
first to apprehend, then comprehend
the significant potentialities
of generalized principles
permeating their physical experiences.
Thus were humans gifted
imaginatively and teleologically
to employ and process
complex information.

Thus also humanity is permitted
by the omni-intellectual,
weightless, amorphous,
metaphysical integrity of Universe —
which we intuitively designate
by the sound word ''god''
to participate in meager degree
locally and temporarily —

in god's own vast
evolutionary designing capabilities.

And of all the designs
thus far formulated by humans
none have been
as adequately anticipatory
of the probable reoccurrences
of yesterday's experiences —
positive and negative,
large and small,
frequent and infrequent,
sudden and slow —
and therefore as
progressively comprehensive
complexedly adequate,
economically exquisite,
powerfully eloquent
and regeneratively reinspiring
to further evolutionary perfection
as is
the sailing ship.

It is visually obvious
even to the inexperienced viewer
that the sailing ship is designed
to cope with nature's
most formidably hostile
environmental conditions
for human survival,
those existing at the interface
of the ocean's and the atmosphere's
ofttimes tumultuous ferocity,

where, for long and most often,
of all places around Earth
an unprepared, ill-equipped humanity
usually perished.

For it was the many lethal experiences
with those myriad
of awesomely demanding conditions
witnessed by a few fortunate survivors
which progressively invoked
man's subjective discovery
and objective invention
of general engineering principles
as well as the foundations of mathematics
from which in turn he evolved
not only competent naval architecture
but such other mathematical essentials as
chronometers, compasses, charts,
spherical trigonometry, sextants
and celestial navigation,
and thereby derived
instrumentally guidable safe passaging
of multitonned vessels
scudding along under full sail
over the rocks-and-shoals-permeated
great ocean waters
under the invisible conditions
of night, fog and high seas.

And sailing ships
unlike bulldozers
do no damage to the sea, land or sky
while employing the windpower

without any depletion
of the vast wealth of universal energy.

And because the sailing ship's beauty
is the unpremeditated consequence
of omni-integrity in designing
both its comprehensively anticipatory performance as a ship
as well as the technology of its building,
that functional beauty has inspired
the high-seas sailorman,
voyaging safely within its womblike hold,
not only reproductively to proliferate
the successful prototype designs
but also spontaneously to identify
sailing ships
as females.

Three quarters of our planet
is covered by water.
And in developing the ability
to live at sea
and thereby to integrate
the world around occurring
but very unevenly distributed
gamut of physical resources and knowledge,
and thus ultimately to make all resources
available to the integrated production
and distributive service of all humanity
(despite the world-around recurrent formidable conditions),
humanity has manifested
its greatest comprehensively anticipatory
scientific designing effectiveness
in the high-seas sailing ship,

the by-products of which have been
establishment of a science-founded,
world-embracing,
scientifically laboratoried,
search and research navigated,
speed of light intercommunicated,
industrial mass-production complex —
out of which, in turn,
has come, evolutionarily,
humanity's mastery of sky and interplanetary travel
and its biochemical conquest
of physiological disorders
of the human organism
and possibly soon to come
the adequate physical sustenance of all mankind.

Key to humanity's scientific discoveries,
technical inventions,
design conceptioning
and production realizations
has been a phenomenon
transcendental to humanity's
self-disciplined
objective concentrations of thought
and deliberate acts —
a phenomenon transcendental to humanity's
consciously disciplined inventive capabilities.

That key is the first
and utterly unpremeditated event
in all discovery, invention and art.
It is humanity's *intuitive* awareness
of having come unwittingly upon

an heretofore unknown truth,
a lucidly conceptual,
sublimely harmonic,
regenerative relationship
of a priori Universe —
an eternal principle —
and then moments later
a second *intuitive* awareness
regarding what the conceiving individual human
must do at once
to capture the awareness of
and secure the usefulness of
that eternally reliable generalized principle
for all humanity
for now and henceforth.

Again and again,
step by step,
intuition opens the doors
that lead to designing
more advantageous rearrangements
of the physical complex of events
which we speak of as the environment,
whose evolutionary transition ever leads
toward the physical and metaphysical success
of all humanity.

And because its design
permits humanity to live anywhere
around our planet's watery mantle
and because this sailing craft
we are now to launch
is the epitome of design competence —

as manifest at this moment
in the forever forwardly mounting and cresting wave
of design capability —
we herewith give
to this world-around dwellable
high-seas sailing craft
the name — INTUITION .

During pauses in the post-launching events
the soliloquy persisted
and later that evening
and for many days thereafter
the following thoughts were inscribed:

I am now seventy-three years of age
and am eager to participate further
in humanity's designing functions —
that is, in metaphysically comprehending
and mastering in orderly ways
the physical energy Universe's
inexorably expanding momentary disorders,
and am aware that humanity
is approaching a crisis
in which its residual ignorance, shortsightedness
and circumstance-biased viewpoints
may dominate,
thus carrying humanity
beyond the "point of no return" —
enveloping this exclusively Sun-regenerated
planetary home
in chain-reactive pollutionings
and utter disorder.

As a comprehensive and anticipatory design scientist
I am aware that the reciprocating engines
of all our automobiles
are only about
fifteen percent efficient,
while our gas turbines
are about thirty,
and our jet engines
about sixty percent efficient,
and fuel cells eighty percent
in delivering effective work power
from the energies they consume.

The overall average of mechanical efficiency
of world-around humanity's power-to-work
as presently designed and tooled-up
is only about four percent,
while experienced engineers and scientists concede
that the world's industrial network
could easily be redesigned to operate at better
than an overall fifteen percent efficiency.

Ergo, I have long been intuitively aware
and am now scientifically confident
that a physically permitted design revolution
is indeed feasible
which can increase fourfold
the present design tool-up
of the planet Earth's technology,
thus raising it to a meager
sixteen percent
overall efficiency,
which can do so very much more
with progressively ever less

of kilowatts, minutes, grams and pollution
of the physical resources of our Spaceship Earth —
to be invested in
each function accomplished
as to be able to raise
the overall percentage of humanity
enjoying a satisfactorily adequate standard of living
to a one hundred percent "haveness,"
and do so
without having any human
prosper at the expense of another
and do all the foregoing
not only for all the humans now aboard
but also for all those
later to come aboard
our Spaceship Earth,
which witnessed a condition
only two thirds of a century ago
when less than one percent
of its human passengers
enjoyed an in any way comparable standard of living.

I am also acutely aware
that only a very small percentage of humanity
has enough comprehensive experience
and cerebrated reconsideration of those experiences
to know that all the foregoing is true.

And if you do not know
all of the foregoing to be true
you do not know enough
to be able to
comprehend the synergetic significance
of the integrated truth,

as well as feel intuitively
the irresistible compulsion
to act effectively
in the teleologic solution
of all those problems,
and thereafter to reduce them
by design science
to realized technological practice
and industrial adoption
at the earliest moment
whereby humankind may be
streamlined into unself-conscious adoption
of ever more effective
new ways of behaving,
thus also unconsciously to abandon
the inadequate customs.

I also realize intuitively
that the elimination
of the condition of resource inadequacy
and thereby the elimination of human want
may probably eliminate war
— or *quick death* —
which is always consequent to the overlong protraction
of the slow and more anguished poverty's
slow dying
as brought about by lethal ignorance
in respect to the design revolution potentials
as society takes its only known recourse
in political actions,
which can but throw the "Ins" out
or "pull the top down,"
unwitting that the design revolution
could effectively elevate

not only all those now on bottom
but also those now already prospering,
bringing all humans alike
to higher levels of advanced living
than have as yet been realized
by any humans,
without taking away
or diminishing the advantages of any.

For the norm of all yesteryears
was failure
as unwillingly conceded
by a sometimes
vainly boastful
but most often abjectly prayerful
poverty-and disease-bewildered people
living out only one third
of their potential years
in utter ignorance
of the invisibly bounteous life-support system
hidden in the superficial landscape,
and consisting only
of instrumentally gleanable information,
abstract and weightless generalized principles,
unique electromagnetic frequencies
and exclusively mathematical realizabilities;
while the norm of today and tomorrow,
if any is possible,
must be total success
for all of humanity
as inherent in
the integratable potentials
of the comprehensive family
of omni-interaccommodative,

and omniorderly
generalized principles
discovered by scientists
to be in a priori governance
of universal evolution's aggregate
of nonsimultaneous
and only partially overlapping
transformative events.

And the norm of sustainable success
of all humanity
will be realized
by the computer-confirmable information
that humanity can afford
to gratify handsomely
whatever of its needs
and growth requirements
can be satisfied
by what can be produced
out of the as yet untapped resources
employed in yesterday's
now obsolete and scrapped
technological devices.

For we now know scientifically
that wealth consists exclusively
of physical energy
which cannot be depleted
plus intelligence's *know-how*
which can only increase
as the ever metaphysically improved uses
of the family
of unique physical behaviors

of energetic Universe
are ever more promptly
reinvested to omniregenerative advantages
of all humanity's ecological involvements
within an omniconsiderate
universal evolution integrity.

But as of this old-to-new era's
threshold crossing moment,
ignorance of the design revolution potentials is pervasive,
and its vacuum persuades
the most powerful political thought
of the largest organized groups of society
— among the sixty percent of humanity
now aboard Earth
who are as yet "have nots" —
to assume that
since there seemingly is nowhere nearly enough
of vital resources
for all to be successful,
and in current fact
only enough to support a minority,
the only fair condition for society
is one of comprehensive deprivation.

And a camaraderie of poverty
which ever and again
must assuage its emotional depression
by vindictively leveling
all attempts of any individual humans
to advance standards
as mistakenly constituting new upshoots
of the socially abhorred

survival only
of the fittest selfishness.

We are also aware
that other vast numbers of the ''have nots''
who are almost entirely
unorganized politically,
have for so many millennia
suffered intensely
both physically and metaphysically
throughout their short-termed lives,
that there has been
no suggestion in their experience
that life was meant to be
anything other than a tortuous trial.

Ergo, they rationalize
that the only explanation
of such a negative experience
that could be hopefully contemplated
is that life on Earth constitutes
only a period of qualification
for an eternal life
hereafter and elsewhere,
and the greater the hardship endured
in the temporary or temporal life
the pleasanter the life hereafter.
And to all such life-hereafterers
any attempt to ameliorate and improve
their short life on Earth
assumedly threatens
to dissipate and preclude

the eternal ecstasy
of their life hereafter.

I am also aware that
an increasing number of human beings —
more especially the young people of the world —
who have witnessed the success
of my earlier prognostications, inventions
and developmental initiatives
in doing more with less
to inaugurate the design revolution
are asking me
at an increasing frequency
and in increasing numbers
to share with them the experimental knowledge
which I have gained
and the philosophically harvested
design strategies I employ.
They seem especially interested
in my environment control devices
which employ only the high priority
sea and sky sciences and technologies
of advanced industrialization.

But most of all they are interested
in synergetics —
which is the name I have given
to the omnirational, comprehensive
co-ordinate system of Universe
which it has been my privilege
to have discovered.
Synergetics makes nuclear physics

a conceptual facility
comprehensible
by any physically normal child.

It also seems clear
that an increasing number of young,
or young-minded people
are beginning
to share my awareness
that total holocaust
is now being ignorantly induced
by the world's preoccupation with
exclusively political palliatives
which are inherently shortsighted
and applicable only
to the emergency-dramatized local aspects
of the greater and unrecognized
evolutionary problems
with which human life
aboard our planet is beset.

But evolution is apparently intent
that life in Universe
must survive.

Biological life
is syntropic
because it sorts and selects
unique chemical elements
from their randomly received
time and locality of receptioning
as celestial imports;
or from out of their random occurrence
as terrestrial resources — fresh or waste —

anywhere around our Earth's biosphere,
and reassociates those elements
in orderly molecular structures
or as orderly organs
of ever-increasing magnitude,
thus effectively reversing
the entropic behaviors
of purely physical phenomena
which give off energy
in ever more random,
expansive and disorderly ways.
For human life contains the weightless
omnipowerful, omniknowing
metaphysical intellect
which alone can comprehend,
sort out, select,
integrate, co-ordinate and cohere.

Little humans
preoccupied with the immediate needs
of their physical regeneration
have locked
their zoom-lens focusing mechanism
on the close-ups only
leaving it exclusively to their intuition
to remind them
once and again in a surprised while
of the vast long-distance focusing
of evolutionary events.

And because evolution is apparently intent
upon accomplishing humanity's total economic success,
whenever society delays overlong
in adopting, producing, distributing and using

in peaceful spontaneity,
the evolutionary essential
discoveries and inventions
of the technological innovations,
evolution then forces humanity to adopt and develop
all the progressively advancing technologies
as emergency commitments
under the negative aegis of group fear of military defeat
and its consequent defensive action taking.

In these emergencies humanity reorganizes
the physical environment
in naturally permitted ways
which turn energy as matter
into a myriad of wheel-mounted levers
and shunt energy
as radiation-induced flows
to impinge upon those levers,
thereby to do the gamut of tasks
conceived by the human mind
to be most productively efficient
and requisite to the immediate survival emergencies,
thereby inducing humanity's
inadvertent acquisition
of the subsequently and peacefully employable
mass-production capability,
which could have been acquired
at fractions of the cost in lives and goods
had they been undertaken peacefully
at the time that they were introduced
by the intuitive inventors, scientists, artists.

Though it was readily discernible long before the war
it was only publicly acknowledged after the war

that the copper mined, refined and shaped into wire
in the emergency
did not unrefine itself
and return as ore into the mountain
after the war was over —
but remained in the dynamo winding
and in the high-voltage transmission lines,
to keep on delivering
electrically converted water power
to distant places
to help humanity, for instance, do its work,
and refrigerate the foods that used to perish
before reaching the world's mouths
thus to regenerate life.
Man had simply rearranged the inanimate scenery
to support more humans
for more days of their lives.
But this cosmically permitted
rearrangement of our planet's
environmental furniture of chemical elements
was erroneously assessed and costed by America
as constituting a vast national indebtedness
and entered only negatively
into the ledgers
of the ignorantly perpetuated,
exclusively depreciative agricultural accounting system.
This historically honored financial system
had been appropriate only
to the inherently perishable
short-term energy conservations
and ecological energy exchangings of bio-organics
accomplished exclusively by photosynthetic impoundment
on planet Earth, of sun and star radiation.
Agricultural economics accounts only

the strictly physical, short-term realizabilities.
Agricultural metabolics differ from industrial metabolics
which deal exclusively with the eternal metaphysical principles
impersonally governing the external, detached processes
of the inherently imperishable, forever regenerative,
physical energy intertransformings of cosmic evolution,
whose inexhaustible inventory of unique capabilities
human minds may employ to produce
progressively amplifying human life support
with ever less units of time, weight, and effort
per each accomplished function.

This metaphysical experience always and only
multiplies irreversibly
as human ''know-how''; whereby, for instance,
the telephone invention and its ever-evolving technology
first conceived of and realized
the transmission of only one conversation
over a given cross section
of imperishable copper wire;
whereafter in seventy-five years of development
mind learned progressively how to conduct
increasing numbers of privately isolated
overlappingly transmitted, individual conversations
over the same cross section of a single wire;
first accomplishing two, then twelve,
then twenty-eight, followed successively
within a few years by ever-multiplying steps
of two hundred and two thousand conversations
over the same original cross section of wire,
and then went wireless!
After which the distance range
of transmittability

and fidelity of sound were amplified
until the 1972 completion
of a world-girdling set of fixedly oriented
communications satellites
which provide world-around high fidelity,
twenty-four-thousand-mile telephonic communications
wherein each one-quarter-ton satellite
is outperforming the transmitting and fidelity capabilities
of one hundred and seventy thousand tons
of transoceanic, bottom laid, copper cables.
The self-regenerative electronics know-how
continually remelted and reworked World War One's
limited North American and European copper inventory
of the world telephone industry
to accomplish the omniplanetary interlinkage of all humanity
while all the time reducing
the total tonnage of copper involved in 1918.
Despite such evidence
of irreversible vantage gain
for every successive reinvestment
of the concomitant service know-how increase
regarding the ninety-two regenerative chemical elements'
infinite reworkability,
the reinvestments are financially capitalized
only as interest-bearing debts and earning obligations.

Initially funded only by cold war's ''emergency'' enactments,
this misaccounting is perpetuated
through politically maneuvered misinterpretation
as constituting only colossal monetary war ''expenditures''
whereas the only true war ''expenditures'' have been
of hours of human life
or of the human lives themselves.

Those human-life expenditures
over-prepaid all costs for all times —
billionsfold.
The only indebtedness
is of human gratitude
to all humans of all times.
Thus all of humanity's common wealth —
its real wealth —
is its established production
and distribution capabilities
as on-going life supports,
both metaphysical and physical,
which have been forever amplified
and will be ever more so — forever.

In view of all the foregoing
it is equally evident
that whatever we can contribute
as individuals
which might lead to humanity's
choosing to abandon with sufficient alacrity
its futile preoccupation with politics —
and the latter's inevitable recourse only to war
and the latter's negatively accounted "spending" —
is dependent upon
our continuing physical health and agility,
metaphysical clarity and
spontaneous initiative.

For if we can maintain
both physical and metaphysical health
we might be able to join with others
in helping to tip the scales

in favor of world society's becoming preoccupied
with the design-revolution priority
and its inherently required
educational-process revolution.

It was thus that intuition
suggested the sloop INTUITION
as the most favorable tool
for self-effectiveness conditioning
within our individual ken.
For ships, sailors and the sea
have been my greatest
teachers and conditioners.

The Ancient Greeks initiated problem solving
by recourse to cosmology and cosmogony,
by proceeding from the whole to the part
lest they miss
the exquisite relevance
of each little part or event.

Thus did the Ionian Greeks intuitively
commence mathematical pattern mensuration
of their world, by geo-metry,
within which comprehensive, synergetic advantage
they discovered and demonstrated that
the known sum —
one hundred and eighty angular degrees —
of all the angles of any triangle,
plus the known values
of three of the triangle's six parts
provided the mathematical capability
to discover the other originally unknown values.

Thus also synergetically did Democritus,
starting with the totally known complex
of visible Universe behaviors,
come to conceive schematically
of the logically necessary existence
of primary yet invisible components
of the physical Universe
which he named ''atoms,''
more than two millennia in advance
of nonsynergetically plodding science's
physical verification
of the microcosmic stardom role
played by those atoms.

If all humanity attains planetary success,
central to that attainment will be
the magnificently regenerative power
of the Greek's intuitive
synergetic spontaneity of thought.

MISTAKE MYSTIQUE

Mistake Mystique

What do you think is the greatest challenge facing young people today as they prepare to assume their caretakership of this world?'' was the question recently asked of me by a midwest high school.

From my viewpoint, by far the greatest challenge facing the young people today is that of responding and conforming only to their own most delicately insistent intuitive awarenesses of what the truth seems to them to be as based on their own experiences and not on what others have interpreted to be the truth regarding events of which neither they nor others have experience-based knowledge.

This also means not yielding unthinkingly to "in" movements or to crowd psychology. This involves assessing thoughtfully one's own urges. It involves understanding but not being swayed by the spontaneous group spirit of youth. It involves thinking before acting in every instance. It involves eschewing all loyalties to other than the truth and love through which the cosmic integrity and absolute wisdom we identify inadequately by the name "God" speaks to each of us directly — and speaks only through our individual awareness of truth and our most spontaneous and powerful emotions of love and compassion.

The whole complex of omni-interaccommodative generalized principles thus far found by science to be governing all the behaviors of universe altogether manifest an infallible wisdom's interconsiderate, unified design, ergo an a priori, intellectual integrity conceptioning, as well as a human intellect discoverability.

This is why youth's self-preparation for planetary caretakership involves commitment to comprehensive concern only with all humanity's welfaring; all the experimentally demonstrable, mathematically generalized principles thus far discovered by humans, and all the special case truths as we progressively discover them — the universally favorable synergetic consequences of which integrating commitments, unpredictable by any of those commitments when they are considered only separately, may well raise the curtain on a new and universally propitious era of humans in universe.

By cosmic designing wisdom we are all born naked, helpless for months, and though superbly equipped cerebrally, utterly lacking in experience, ergo utterly ignorant. We were also endowed with hunger, thirst, curiosity, and procreative urge. We were designed predominantly of water — which freezes, boils, and evaporates within a miniscule temperature range. The brain's information-apprehending, — storing, and — retrieving functions, as the control centers of the physical organisms employed by our metaphysical minds, were altogether designed to prosper initially only within those close thermal and other biospheric limits of planet Earth.

Under all the foregoing conditions, whatever humans have learned had to be learned as a consequence only of trial-and-error experience. Humans have learned only through mistakes. The billions of humans in history have had to make quadrillions of mistakes to have arrived at the state where we now have 150,000 common words to identify that many unique and only metaphysically comprehensible nuances of experience. The number of words in the dictionary will always multiply as we experience the progressive complex of cosmic episodes of Scenario Universe, making many new mistakes within the new set of unfamiliar circumstances. This provokes thoughtful reconsideration, and determination to avoid future mistake making under these latest given circumstances. This in turn occasions the inventing of more incisively effective word tools to cope with the newly familiar phenomena.

Also by wisdom of the great design, humans have the capability to formulate and communicate from generation to generation their newly evolved thoughts regarding these lessons of greater experience which are only expressable through those new words and thus progressively to accumulate new knowledge, new viewpoints, and new wisdom, by sharing the exclusively self-discovered significance of the new nuances of thought.

Those quadrillions of mistakes were the price paid by humanity for its surprising competence as presently accrued synergetically, for the first time in history, to cope successfully on behalf of all humanity with all problems of physically healthy survival, enlightening growth, and initiative accommodation.

Chagrin and mortification caused by their progressively self-discovered quadrillions of errors would long ago have given humanity such an inferiority complex that it would have become too discouraged to continue with the life experience. To avoid such a proclivity, humans were designedly given pride, vanity, and inventive memory, which, all together, can and usually do incline us to self-deception.

Witnessing the mistakes of others, the preconditioned crowd, reflexing, says, ''Why did that individual make such a stupid mistake? We knew the answer all the time.'' So effective has been the nonthinking, group deceit of humanity that it now says, ''Nobody should make mistakes,'' and punishes people for making mistakes. In love-generated fear for their children's future life in days beyond their own survival, parents train their children to avoid making mistakes lest they be put at a social disadvantage.

Thus humanity has developed a comprehensive, mutual self-deception and has made the total mistake of not perceiving that realistic thinking accrues only after mistake making, which is the cosmic wisdom's most cogent way of teaching each of us how to carry on. It is only at the moment of humans' realistic admission to selves of having made a mistake that they are closest to that mysterious integrity governing the universe. Only then are humans able to free themselves of the misconceptions that have brought about their mistakes. With the misconceptions out of the way, they have their first view of the truth and immediately subsequent insights into the significance of the misconception as usually fostered by their pride and vanity, or by unthinking popular accord.

The courage to adhere to the truth as we learn it involves, then, the courage to face ourselves with the clear admission of all the mistakes we have made. Mistakes are sins only when not admitted. Etymologically, sin means *omission* where *admission* should have occurred. An angle is a *sinus,* an opening, a break in a circle, an omission in the ever-evolving integrity of the whole human individual. Trigonometrically, the sine of an angle is the ratio of the length of the side facing the central angle considered, as ratioed to the length of the radius of the circle.

Human beings were given a left foot and a right foot to make a mistake first to the left, then to the right, left again, and repeat. Between the over-controlled steering impulses, humans inadvertently attain the between-the-two desired direction of advance. This is not only the way humans work — it is the way the universe works. This is why physics has found no straight lines; it has found a physical universe consisting only of waves.

Cybernetics, the Greek word for the steering of a boat, was first employed by Norbert Weiner to identify the human process of gaining and employing information. When a rudder of a ship of either the air or sea is angled to one side or the other of the ship's keel line, the ship's hull begins to rotate around its pivot point. The momentum of that pivoting tends to keep rotating the ship beyond the helmsman's intention. He or she therefore has to "meet" that course-altering momentum whose momentum in turn has again to be met. It is impossible to eliminate altogether the ship's course realterations. It is possible only to reduce the degree of successive angular errors by ever more sensitive, frequent, and gentle corrections. That's what good helmsmen or good airplane pilots do.

Norbert Weiner next invented the word *feedback* to identify discovery of all such biased errors and the mechanism of their over-corrections. In such angular error correction systems (as governed, for instance, by the true north-holding direction sustained by the powerful angular momentum of gyroscopes which are connected by delicate hydro- or electrically actuated servo-mechanisms to the powerful rudder-steering motors), the magnitude of rightward and leftward veering is significantly reduced. Such automated steering is accomplished only by minimizing angular errors, and not by eliminating them, and certainly not by pretending they do not exist. Gyro-steering produces a wavilinear course, with errors of much higher frequency of alternate correction and of much lesser wave depth than those made by the human handling of the rudder.

All designing of the universe is accomplished only through such alternating angle and frequency modulation. The DNA-RNA codes found within the protein shells of viruses which govern the designing of all known terrestrial species of biological organisms consist only of angle and frequency modulating instructions.

At present, teachers, professors, and their helpers go over the students' examinations, looking for errors. They usually ratio the percentage of error to the percentage of correctly remembered concepts to which the students have been exposed. I suggest that the teaching world alter this practice and adopt the requirement that all students periodically submit a written account of all the mistakes they have made, not only regarding the course subject, but in their self-discipline during the term, while also recording what they have learned from the recognition that they have made the mistakes; the reports should summarize what it is they have really learned, not only in their courses, but on their own intuition and initiative. I suggest, then, that the faculty be marked as well as the students on a basis of their effectiveness in helping the students to learn anything important about any subject — doing so by nature's prescribed trial and error leverage. The more mistakes the students discover, the higher their grade.

The greatest lesson that nature is now trying to teach humanity is that when the bumblebee goes after its honey, it inadvertently pollinizes the vegetation, which pollinization, accomplished at 90 degrees to the bumblebee's aimed activity, constitutes part of the link-up of the great ecological regeneration of the capability of terrestrial vegetation to impound upon our planet enough of the sun's radiation energy to support regeneration of life on our planet, possibly in turn to support the continuation of humans, whose minds are uniquely capable of discovering some of the eternal laws of universe and thereby to serve as local universe problem solvers in local maintenance of the integrity of eternal regeneration of the universe.

In the same indirect way, humanity is at present being taught by nature that its armament making as a way to make a living for itself is inadvertently producing side effects of gained knowledge of how to do ever more with ever less and how, therewith, to render all the resources on earth capable of successful support of all humanity. The big lesson, then, is called *precession*. The 90 degree precessional resultants of the interaction of forces in universe teach humanity that what it thought were the side effects are the main effects, and vice versa.

What, then, are the side effects of knowledge gained by students as a consequence of the teacher's attempt to focus the students' attention on single subjects? It can be that all the categories of informational educational systems' studies are like the honey-bearing flowers, and that the really important consequence of the educational system is not the special case information that the students gain from any special subject, but the side effects learning of the interrelatedness of all things — and thereby the individual personal discovery of an overall sense of the omnipresence and reliability of generalized principles governing the omnirelatedness — whereby, in turn, the individuals discover their own cosmic significance as co-functions of the "otherness," which co-functioning is first responsible to all others (not self), and to the truth which is God, which embraces and permeates Scenario Universe.

The motto of Milton Academy, the Harvard preparatory school I attended, was "Dare to Be True." In the crowd psychology and mores of that pre-World War I period, the students interpreted this motto as a challenge rather than an admonition, ergo, as: "Dare to tell the truth as you see it and you'll find yourself in trouble. Better to learn how the story goes that everybody accepts and stick with that."

Ralph Waldo Emerson said, "Poetry means saying the most important things in the simplest way." I might have answered the school in a much more poetical way by quoting only the motto of 340-year-old Harvard University, "*Veritas* (Vere-i-tas), meaning progressively minimizing the magnitude of our veering to one side or the other of the star by which we steer, whose pathway to us is delicately reflected on the sea of life, and along whose twinkling stepping-stone path we attempt to travel toward that which is God — toward truth so exquisite to be dimensionless, yet from moment to moment so reinformative as to guarantee the integrity of eternally regenerative Scenario Universe.

Veritas — it will never be superseded.

BRAIN AND MIND

R. BUCKMINSTER FULLER

Brain and Mind

Dr. Harvey Cushing — 1869 to 1939 —
was so great a neurosurgeon
that his professional colleagues first called themselves
The Harvey Cushing Society
but later adopted the more formal name
of American Association of Neurosurgeons
and at the same time instituted the Harvey Cushing Oration
as the principal address of their annual congress.
And though I am neither a neurosurgeon
nor a professional of any discipline
an aberration of fate brought me the honor of delivering
''The 1967 Harvey Cushing Oration''
to two thousand of their members
at their annual meeting in Chicago.

I never prepare lectures so I thought out loud to them
about humanity, its world
and its function in universal evolution,
and as I thought and spoke I realized o'erwhelmingly
that if humanity is going to survive
it will be only because it commits itself
unselfishly, courageously and exclusively
to its most longingfully creative inclinations,
visionary conceptions
and intuitively formulated objectifications;
for the mind's intellections —
in contradistinction to the brain's automatics —
apparently constitute humanity's
last and highest order of survival recourse.

Therefore I felt I must devote the occasion
to distinguishing between mind and brain —
for that unique difference
also differentiates most incisively
between human beings
and all the other living creatures.
For instance, it was mind alone that discerned
that physical experiment disclosed and confirmed
that local physical systems
are always exporting energy
in one manner or another, such as by friction,
and it was mind alone that determined to identify semantically
the exporting of energy
by inventing the abstract name *entropy*.

And mind went on to discern
that physical experiences disclosed and confirmed
that all living tissue
during cell multiplication
must import more energy
than it exports,
else it could neither grow
nor even sustain healthy balance.
And mind also witnessed
that crystalline structures
also can import energy
but not as much as they export;
and mind identified
energy importing by the name *syntropy*.

Because of the tidal fluctuations of syntropy-entropy
local environments are forever altering themselves.
Living phenomena, being both entropic and antientropic,
are, as Professor Waddington points out,
forever altering the environment
at a faster rate than the nonbiologicals,
and the ever-more-completely altered environment
is, in turn, continually altering all the creatures.
Waddington identified this external modification
of living morphology as epigenetics —
in contradistinction to the corporeal morphology
of all living organisms' integral growth
whose angle and frequency designing
is governed by the internal DNA-RNA genetic codes.

As the irreversible succession of self-regenerative human events —
experiences, intuitions, experiments, discoveries and productions —
successively increases
both the comprehensions and capability options,
the commonwealth of intercommunicated comprehensions
produces an ever-evolving, subconsciously changing common sense.
Where syntropy is gaining over entropy, life prevails;
where entropy is gaining over syntropy, death prevails.
Their exponentially regenerative, birth-death interplay
is describable in information theory
as "self-accelerating feedback,"
and in nuclear physics it is manifest as "chain reaction."
And in an even more comprehensive way
it is manifest pulsatingly, resonantly and propagatively,
as the irreversible regeneration of universal evolution.

For the obviously inanimate,
nonbiological, physical phenomena
are all, always, giving off energies
in ever more diffuse, expansive
and disorderly ways,
which impose complex intertransactions
upon all the intertransforming systems.
This was only half anticipated
in a generalized way
by the late eighteenth-century scientists'
academically hailed
great ''Second Law of Thermodynamics,''
which discovered
and recognized only
the energy-exporting phase called entropy.

Entropy's behavior may be modernized to state
that every separately experienceable
and generalizably conceivable system in Universe
is continually exporting energies
while also always importing energies
at a concurrently accelerating and decelerating
variety of local system rates,
which also means
that all systems are continually transforming
internally as well as externally,
and because the periodicity of importing and exporting
are both nonsimultaneous and unequal,
all the systems are tidally pulsative
at a variety of frequencies.

In the same way that systems
have "centers of gravity" (CG)
and "neutral axes of gyration"
identified by engineers as "I,"
they also have
"centers of omniequilibrious symmetry,"
at which their kinetic transformings never pause,
but relative to which kinetic action centers
they oscillatingly transform.
And the frequencies and geometrics
of those internally-externally co-ordinated pulsative trendings
are always uniquely asymmetric
as related to the local systems' symmetrically co-ordinate
abstract "centers of equilibrious symmetry" (CES).

Because the unsynchronizable, asymmetric excesses
are inherently exported,
the internal-external events
propagate both inward-and outward-bound waves.
These unique wave-system propagations
only infrequently coincide
with the unique symmetry patternings of others.
The orderly patterning energy releases
of any one system
only superficially appear to be disorderly —
being unsynchronized immediately with other systems,
though each system is internally orderly
and each is uniquely symmetrical dynamically.
This relatively minor yet true disorder, external to local systems,
is spoken of by the confused observer as ''diffuse.''

Having different-sized teeth, and rates of revolution,
two such gears cannot mesh, but associate only tangentially.
Consequently, their axial centers must be farther apart
than are those of meshable gears.
Omnidirectionally pulsative systems
are, in effect, spherical gears.
Their inwardly and outwardly pulsating and rotating "teeth"
consist of multifrequenced circumferential and radial waves
of fifty-six great-circle subdivisions of spherical unity,
often nonmeshing with other local systems.
The universally infrequent meshing
of wavelengths and frequencies produces an omnicondition
in which the new omnidirectional system's center must,
as each is created,
continually occupy omnidirectionally greater domains of disorder.

The sum total consequence of entropy is
an omniexpanding physical Universe
and an only (apparently increasing) disorderliness.
This does not mean absolute disorder;
it means the momentarily superficial appearance
of less order than symmetry.
And the disorder is only relative
to the majority of individual cases,
for each system
and its particular entropic exportings
is orderly within itself,
and the detection of disorder is mistakenly assumed
as the result of exclusively myopic
and too short-termed observing.

I am convinced therefore
that there is a great deal of difference
between absolute disorder, i.e., chaos,
and the only one-sidedly considered,
relative asymmetry, whose pulsative balancing
at a later time with other systems was not awaited
by the too hasty and biased observer.
On the contrary, I am convinced
by comprehensively considered experience
that a total integrity of order prevails
and am inspired to explore that order
in hope of discovering humanity's function
in the evolutionary scenario
of omni-self-regenerative Universe.

To date, we have gained vast inventories
of trial-and-error experience
from all of which information we have extracted
a family of generalized scientific principles
which are weightless pattern concepts.
Being weightless they are metaphysical.
From the metaphysics
we have in turn designed
rearrangements of the physical behavior constituents
of our omnikinetic environment scenery.
We have rearranged the scenery
in the pattern of world-around occurring
power-driven tool networks
all of which teleologic process
has produced an ever-increasing survival advantage for humanity.

The human advantage is both physical and metaphysical,
as ever-increasing proportions of all Earthians
become involved in the processes
of massive production and distribution
of both the necessities and desirables
which implement
humanity's regenerative evolution requirements.
And the degrees of increasing advantage
are expressible in precise scientific terms
of the number of centimeters, grams and seconds of work
humanity is able to accomplish
out of each hour of investment
of each and all of its individuals'
potential lifetime hours,
energies, materials and know-how.

But we also discover that humanity
does not yet realize its potentially imminent success,
despite doubling and even tripling of life spans
of a billion and a half humans rendered wealthy by industry
all within only the last seventy years,
and despite the concurrent fractionalizing
of human hours individually invested
to produce the essentials and desirables.
Humanity's enlightenment probably is delayed
because the Earth planet is so large,
and humans are so infinitely tiny
and so myopically preoccupied with personally avoiding
the erroneously assumed inevitability of poverty for the many,
which has slavishly and fearfully conditioned their reflexes.

Those not as yet included
in the high-living advantage, ever multiplyingly produced
by power-driven tool networks,
do not comprehend the swiftly accelerating rate
at which comprehensively increasing human advantage
will include them and their children —
as well as the children of the already advantaged —
for they find themselves in a cultural environment
whose customs, logic, and law
were designed uniquely to cope only with the lethal struggling
of the preindustrial, frequently failing agrarian era.
Yesterday's requirement of "earning the right to live"
is contradicted by today's omnisuccess potential.
Children intuit wholesale fallacy in yesterday's imperatives.

Inspired by the magnificent vision
of children's earnest questioning
and in continuance of our attempt
to differentiate the domains and functionings
of brain and mind,
we eagerly explore
for experiential evidence
that human minds and brains
may both be essentials
in the total design integrity
of eternally self-regenerative
Scenario Universe.
From physics we learn that every fundamental behavior
of Universe always and only coexists
with a nonmirror-imaged complementary.

The nonsimultaneity and dissimilarity
of the complementary interpatterning pulsations
integrate to produce
the complex of events
we sensorially identify as reality.
Without the pulsative asymmetries
and asynchronous lags,
the complementations would cancel out one another
and centralize equilibriously,
and there would be no sensoriality,
ergo, no self-awareness, no life.
For we have also learned from physics
that all positive and negative weights
of the fundamental components of matter
balance out exactly as zero.

Life may well be a dream,
a comedy and tragedy of errors of conceptioning
inherent in the dualistic imaginary assumption
of a *self*, differentiated
from all the complex *otherness*,
of reasonably conceivable Universe,
for it must be remembered
that no humans have ever seen directly
outside themselves.
What we call seeing
is the interpretive imagining in the brain
of the significance and meaning
of the nervous system reports
of an assumed outsideness of self.

All of this organic design conception
may be that of a cosmic intellect
which is inventing Universe progressively,
evolving mathematically elegant integral equations
for each conceivable challenge
including the invention you and me.
But you and I cannot escape
and are given extraordinary faculties
which we are supposed to use.
So here we go again
from right where we are *Now*.
At least our speculative excursion
was relative to our attempt to differentiate
between brain and mind.

For the written record of two millenniums
discloses human minds forever rediscovering
the great dream concept.
No waking human has ever proven
it was self that fell earlier asleep.
While the brain of the dog
sleeping at our feet may go dream-hunting,
dogs have never given evidence
of being concerned with metaphysics.
Their brains tell them to wag their tails
to signal "friend" or to bark reflexively
at strange noises. Individual living integrity,
canine or human, may well constitute Soul;
while mind may speculate Soul, Soul is not mind.

Returning to our hard and soft experiential realities,
we seek to discover whether there exists
a major phase of universal behavior
that complements but does not mirror
the omniexpansive, increasing asymmetry
and progressive entropic diffusion of physical systems,
and matches the latter
with an increasingly orderly and symmetrical
omnicontracting and compacting syntropic phase.
We know that the light by which astronomers
are able to observe stars through optical telescopes
is the ''disorderly entropic output'' of radiation
emanating from the ever-transforming
and moving star systems.

Thus, all the information of astronomers until recent decades
has indicated increasing disorder and expansion,
which would seemingly threaten
a progression of Universe
toward ultimate diffusive dissipation
in utter chaos
were it not for Einstein's observation
that the experimentally demonstrated
linear speeds of all forms of radiation
unleashed in a vacuum
are identical,
indicating a top speed
of physical Universe expansiveness
which implies ultimately total cosmic order.

But it did not explain how Humpty Dumpty reassembled himself.
In recent years the astronomers have been aided by radar,
the electronic telescope with which they are potentially capable
of bouncing signals off invisible celestial bodies
which signals could be angled to echo back to Earth,
thus to identify their relative time-fixed celestial positions.
But scientists have not as yet with any certainty
bounced their radar signals off another planet
in a star system other than the Sun's.
Looking for regions or bodies in Universe
where energy events are not only collecting
but doing so with increasing orderliness, that is syntropically,
we discover the cosmically adequate syntropic qualifications
of our own noncandescent spacevehicle Earth.

Humans have continually overlooked Earth
in their celestial searching
for syntropic bodies
because of their fundamental propensity
for not thinking realistically of the Earth
as an astronomical body.
Using a miniature globe to demonstrate
that our Earth is a sphere
they conceded theoretically
that Earth is not an infinite plane,
but they as yet feel and talk
of ''the four corners of the Earth,''
of ''the wide, wide world,''
and identify ''realistic, practical'' thinking
as ''getting down to Earth.'' Out of space!

They use the words *up* and *down*,
which refer exclusively to a planar concept
of the world and Universe;
for all the perpendiculars to an infinite plane
must be parallel to one another,
ergo, extend only upwards and downwards,
"up to the Sky," "down to Earth."
Because humans see illusionarily
that receding parallel tracks apparently converge
before reaching the horizon,
they long, long ago misassumed
that all parallel lines leading "upwardly"
must also converge in a point called *Heaven*
and converge in the opposite direction
in a spot called *Hell*.

This illusion as yet permeates Earthians' reflexing.
Even though many humans no longer promulgate
the conceptions of an otherworldly Heaven and Hell,
they retain the misorientation of Up and Down.
I have asked hundreds of audiences around the world
for a show of hands by those
who do not use the words "up" and "down."
In none of my audiences have hands appeared.
This means that all the human beings
in all of my audiences
use the words "up" and "down," and in so doing
believe themselves eminently logical.
And they query, "What may we say cogently
in lieu of *up* and *down?*"

The answer was found by aviators.
Having flown half around the world,
they did not feel "upside down."
As flyers they selected the right words.
Flyers "Come *in* for a landing."
Flyers "Go *out* to the sky."
Flyers "Fly *around* the world."
Astronauts go *in* toward various celestial bodies
and accelerate *out*wardly from them
and *into* the spatial nothingness
out of which they steer themselves
to come *in* to another
of the always orbiting celestial bodies,
around any of which they may locally orbit.

All of us go *out*ward, *in*ward, and *around*
any object, or system of reference,
such as planets, stars, houses, things, and atoms.
Atomic events transpire
*in*ward, *out*ward, and *around* their respective nuclei.
We direct astronauts *into* the Moon:
we will soon direct them *into* other planets.
Out is common to all bodies
and is *out*ward in all directions
from any one of them.
"Out" is omnidirectional,
"In" is unidirectional. *In* is unique.
In is always specifically oriented.
In is *in*dividual, specific. *Out* is any way to nowhere.

The word "invention"
uses the prefix "in"
to identify specifically
a coming into our thought of a unique conception,
which we *in* turn
realize *in* a special physical case demonstration
thus *in*-troducing
the *in*-vention to society.
No scientist would suggest that any locus of the Universe
is identifiable as "up"
nor any other as "down."
Yet individual scientists themselves are sunsets
and as yet reflex spontaneously
in an "up" and "down" conditioned miasma.

Their senses say reflexively,
"I see the Sun is going down,"
despite that scientists have known theoretically
for five hundred years that the Sun is not
going down at dusk and rising at dawn.
What is critically significant in this connection
is the way in which humans do *reflex spontaneously*
for that is the way in which
they behave in surprise crises. Humans often
find it expedient to yield reflexively
in perversely ignorant practical ways
which culminate in social disasters
that deny the significance of theoretical knowledge.
We often yield spontaneously to fallacious *common sense*.

We are therefore interested in how we humans,
can liberate ourselves
from all the false up and down reflexing.
I suggest to audiences that they try saying
"I'm going 'outstairs' and 'instairs,'"
which though logical sounds strange to them.
They all laugh about it.
But if they persist saying *in* and *out*
for a few days, even as a joke,
they find themselves beginning to realize
that they are indeed going inward and outward
in respect to the center of our Spaceship Earth.
And for the first time
they begin to feel planetary.

Blocked until now by the *heavens above*
and *down to earth* fixations,
humanity has failed to realize commonsensically
that our own Spaceship Earth
is an independent celestial object,
albeit a swiftly moving one,
compared to whose sixty-thousand-miles-per-hour travel
hurrying hurricanes rate as frozen pudding.
Disciplining ourselves to use *in* and *out*
requires frequently thoughtful self-correction
and gradually our knowledge-informed thought
begins to realize realistically
that we are indeed riding
within the thin gaseous skin of a planet.

Studying the experientially acquired data
we begin to discover that energies emanating
from celestial regions remote from planet Earth
are indeed converging and accumulating
in planet Earth's biosphere, top soil, and oceans.
Earth is a spherical importer of energies,
both as radiation and as matter.
Recent estimates of geo- and astrophysicists
show many stardust tons landing daily on Earth.
Some estimate one hundred thousand tons daily —
probably acquired during Earth's orbital passaging
through the rubble of comet trails.
By virtue of such stardust and asteroid fall-ins,
Earth is actually increasing its weight.

But so are the Moon and other planets,
wherefore gravitational imbalance is avoided.
All the stars give off energies
and much of the radiation from stars
other than the Sun
impinge on our Earth.
This cosmic radiation seems to impinge
in the same disorderly manner
as does the stardust, the introverted
concentrate of cosmic radiation.
Every chemical phenomenon can be identified
either by its mass characteristics,
such as weight per volume,
or by its radiation-frequency bands.

Both the frequencies and the matter
are behavioral states of the same phenomenon.
This underlies Einstein's fundamental thinking,
of energy associative as matter (stardust);
or of energy disassociative as radiation;
and of their eternally regenerative
terminal intertransformabilities.
Our particular radiant-energy star — the Sun —
is our prime energy supply source.
The Sun is our nearest celestial fuel ship.
It is flying in formation with us
through the Galactic system
at an Earthian-life's incineration-proofing
distance of ninety-three million miles.

All the while our energy-concentrating
spherical spacevehicle Earth
circles around our ten-billionfold greater
mass-energy, mothership Sun
while co-orbiting our mutual Galactic center —
together with our planetary companions
of the Sun-commanded space fleet —
we aboard Earth are receiving gratis
just the amount of prime energy wealth
to regenerate biological life on board,
despite our manifold ignorance,
concomitant wastage, and pollution.
That Universe tolerates our protracted nonsense
suggests significant unrealized potentials.

Our spherical spacecraft Earth and its passengers
may indeed have a unique cosmic function potential
soon to be realized as our particular star group —
the Sun squadron of planetary spaceships
maintaining its Milky-Way merry-go-round's radial position —
spins Galactically at twelve times Earth's orbital velocity
of sixty thousand miles per hour around the Sun.
Our spiral nebula's one hundred billion co-spinning stars
themselves constitute only a local spacevehicle fleet
in the Grand Fleet of Universe.
While our Galactic fleet maneuvers at cosmic flank-speeds
amongst the billions of other now-known Galaxies
Earth passengers, insensible of their own space travel say
"We don't see how you can *stand* so much traveling."

They are referring, of course, only
to my annually meager 180,000 terrestrial miles
as I jet almost daily or sleep in flight,
or in two hundred different beds
during two hundred different nights of the year
somewhere around our minuscule planet,
and dwell for another hundred nights
in the general facilities of world airlines,
the remaining sixty-five nights of the year
finding me asleep in one of three beds;
one on the East Coast, one on the West Coast,
and the other in the continental middle of North America
of planet Earth which every fifteen minutes
celestially out-travels my annual Earth rounding.

My one hundred and eighty thousand per year
is trivial when compared to an astronaut's
one million miles per week,
or the Moon's six billion miles per year
sun-orbiting corkscrew travel.
The cosmic command-speed of our spacevehicle Earth
is maintained at critical radiational distance from Sun
suitable for biological regeneration.
Also, as protection against the Sun's incinerating us
we have the atomic-hydrogen and Van Allen belts.
These belts are the outermost
of the biosphere's spherical mantles.
Within these atomic-hydrogen and Van Allens we have
the ionic, ozone, and atmospheric spherical veils.

The Van Allen belts intercept the radiation
which would kill a naked man positioned outward of them.
The hydrogen, the Van Allens, the ionosphere,
stratosphere and atmosphere all refractively differentiate
and distribute the radiation frequencies at sub-lethal levels,
separating the original radiation concentrations
into a variety of indirect life-sustaining energy transactions.
Enough Sun light gets through once in a while
to fatally burn naked humans at interface
of the crystalline and hydrosphere levels.
Though it sends no bills
the Sun's radiant energy
is the prime regenerating source
for all biological "life" on our planet.

Even while sunburning their skins
humans and all other mammals
are unable to take in enough radiant energy
to keep them alive.
To circumvent mammals attempting futilely so to do
nature has invented
the green vegetation on the dry lands,
and the algae in the waters around the Earth.
The vegetation and the algae
impound the Sun's radiation by photosynthesis
converting the radiation
into orderly molecules,
which constitute biological life's
prime energy harvest.

The dry land vegetation and water-borne algae
provide metabolic sustenance
of all manner of biological species
some of which energy relayers can in turn
nourish humans directly.
Self-startered by subconsciously initiated desire,
or genetically ''ticker-taped'' urges
to experience thirst, appetite, and breathing,
biological species are subconsciously programmed
to consume the solid, liquid, and gaseous
chemical constituents necessary
to produce the ongoing biological molecules
whose energies are convertible
into protoplasmic tissue growths and muscular actions.

As the prime energy impounder,
the vegetation of the land has to have roots
in order to get enough cooling water
so that it will not be dehydrated
while it photosynthesizes the radiant energy of the Sun
into the beautiful molecular structures
that provide the metabolic energy exchange means
of terrestrial life support.
The algae floating in the sea
are automatically water-cooled.
All this is relevant to our search for an understanding
of Humanity's functioning in Universe
for sumtotally Earth manifests what we first sought,
a moving locus in Universe where syntropy reigns.

Thus we intuit excitedly that
the photosynthesis process
of orderly molecule production
constitutes elegant scientific disclosure
that our planet Earth indeed may be
one such moving locus in Universe
where energy is accumulating syntropically
being thereby conserved in a variety
of ever more compactly and orderly patterned
biological, crystalline,
liquid, and gaseous substances —
as cosmic complementation
of the entropic disorder
ever more myriadly manifest
by the omniexporting star centers of Universe.

As already noted, Number One Manifest
that our planet Earth is just such a syntropic locus
is the constant terrestrial acquisition of energies
around Earth's spherical surface
as provided by both stardust and cosmic radiation.
We note that the cosmic radiation, including the Sun's,
is *not* reflectively redistributed back outward to Universe,
as does a mirrored ball reflectively reject radiation.
Instead, Earth is measurably impounding the radiation
by progressive angular refractions which separate
the originally lethal radiation dosages into nonlethal fractions
and progressively shunt
those frequency differentialized radiations
from perpendicular to circumferential terrestrial travel
within the biosphere's concentric mantles.

This Earthian biosphere's refraction of radiation
manifests mathematically orderly, *angular* sorting
of the Sun's radiation
into separately discrete frequencies.
This is witnessable, for instance, in a rainbow,
or in the twilight sky's
red, orange, yellow, green, blue, and violet
horizontal stratifications.
This *relay system* of *angle* and *frequency modulating*
and biosphere refractioning
constitutes Manifest Number Two
of our sought-for functional identity of Earth
as a syntropic, orderly, energy concentrating,
mobile locus in Universe.

Manifest Three that our Earth is a traveling locus
of syntropic energy concentration in Universe
is the demonstrable fact already noted
that all the biologicals are *continually multiplying*
their orderly cellular, molecular, and atomic structurings
which metabolic conservation functioning completes
the comprehensive pattern integrity equation
governing orderly cosmic energy export-import balancing.
Manifest Four that our planet Earth is surely
the first known syntropic centering of Universe
is its star-dusted, chemically regenerative topsoiling.
Fifth Manifest is the spherical enmantling of biological residues
as hydrocarbons are pressure transformed into coal and petroleum
which as fossil fuels stabilely store the cosmic energy harvest.

Having set out to discover evolutionary experience clues
as to whether humans have an essential cosmic function —
despite misassumption of exclusively self-eminent roles
only as audiences or actors in the Earthian drama ''Life'' —
we sought first to learn whether the Earth planet itself
has its essential function in the Universe. Saying to ourselves
that if Earth's cosmic system function could be found
then we might differentiate out its subsystem functions
thereby to uncover which of Earth's universal functionings
humans might be uniquely performing.
We have thus far found a hierarchy of five Manifests,
clearly confirmatory of planet Earth's functioning
as the only known traveling focus
of syntropic reconcentrations
of the physical energies of eternally regenerative Universe.

This powerfully reinforces our initial assumption
that we had first to find such a syntropic traveling locus
within the total complementary scheme
of universal regeneration.
A Sixth Manifest of Earth's
unique celestial functioning in this syntropic manner
was the "Geophysical-Year" scientists' discovery
of the impoundment of star-energy radiation
in both the Earth's atmosphere and hydrosphere,
which power, temperature,
and pressure the weather and ocean currents
and maintain the critical local microenvironments
within which the biological proliferation of photosynthesis
and subsequent organic transformations occur
as metabolically fed-back chemical exchangings.

For an instance, the heating of the hydrosphere
involves the fact that water takes on heat and loses it
at the slowest known rate of all substances.
As a consequence the temperatures of the watery mantle
covering three-quarters of Earth vary between such close limits
that the world average temperatures throughout the years
have varied less than one degree Fahrenheit
over all the years in which temperatures
have been recorded around the world.
Within these exquisitely stable electro, thermal, chemical limits
the metabolic regeneration of humans is sustained
as the apparently ultimate focal formulation
of the metabolic interchangings and intertransformings
of the total evolution of bio-ecological intercomplementation.

So delicate are the microclimatic-ecological balances
that humans at all times manifest
an internally permeative organic operating temperature
of 98.6 degrees Fahrenheit
no matter what their age,
their geographical location,
or their clothing may be.
Manifest number Seven of Earth's cosmic functioning
is its progressive geological submerging
of the hydrocarbon energy residue concentrates
buried ever more deeply and at increasing pressures
either within the Earth's crust, or its hydrosphere,
whereby those biological residues are chemically transformed
into rigid, liquid, or gaseous fossil fuels.

With these stored physical energies
and the intellectually harvested metaphysical principles
informing experience-generated know-how
humans were able to initiate
the inanimate energy powering of their myriad levers
of industrial tooling,
finally enabling humanity
to employ directly
the inexhaustible energies
of eternally regenerative extraterrestrial Universe,
thus obsoleting humanity's absolute metabolic dependency
upon the frequently failing, lethally limited,
agriculturally precarious energy impoundments,
replaced by direct access to celestially unlimited energy.

Thus Earth's cosmic functions combine synergetically to manifest
that our planet is verily one such energy collecting,
concentrating, sorting, storing, and conserving locus
whose reflectance of star-energy emanations
is greatly exceeded by its cosmic-energies inhibitance.
But Earth's cosmic role as syntropic complementation
of optically obvious entropic star energy disbursements
has remained unconsidered by Earthians, their innate
celestial perceptivity benumbed by down-to-Earthness.
Thus it is evidenced also that Earth's energy concentrating
may culminate multi-millions of years hence
in Earth itself in turn becoming an entropically radiant star
exporting energies for biological life support of human minds
long since migrated to function on planets of other stars.

This is celestial confirmation
of Boltzmann's law,
which states in modernized terms
that within a closed system
there are oscillations and evolutions
between high and low energy
concentrations and diffusions.
New lows concentrate energy and become highs
by exhausting yesterday's highs
as yesterday's exhausted highs
in turn become today's lows.
Boltzmann made his finding
while checking Avogadro's discovery
regarding chemical gas commonalities.

Avogadro intuited, then proved,
that under identical conditions
of energy as heat or pressure
all gases will disclose the same number
of molecules per given volume.
Boltzmann's law has been found to hold true
outside the gaseous microcosm
within which he found it.
It explains for instance
not only the weather —
but also the *high* and *low* energy concentration dynamics
of our biosphere and now we find it explaining
both Earth's and humanity's functioning
in the macrocosmic scheme of Universe.

Thus we come to humans' own unique functioning,
aboard the energy-storing planet Earth,
as distinctly differentiated out
and contrasted to the first seven Manifests
of Earth's cosmic syntropic concentration function
within this moving planetary locus of Universe.
Thus humans' mind-over-mattering distinguishes itself
as cosmic function Manifest Number Eight.
Over and above its syntropic
physical sorting and rearranging
cerebrally reflexed planetary capabilities
we find humanity's metaphysical problem-solving capability
to be uniquely and exclusively referenced
to the complex of eternal principles.

Further confirmation of our apparent discovery
that humanity has
a unique and essential role in Universe
will depend upon our developing a clear understanding
not only of the difference
between brain and mind functioning
but also of their respectively unique characteristics.

Starting a new line of exploration we first note
that until the present moment in history
humanity has not differentiated lucidly
between the meanings of the words
brain and *mind:*
they are often used synonymously.
The pragmatist tends to use
the word *mind* as embracing
what seems ''untenable mysticism'';
while such realists feel also
that the word *brain*
is quite adequate to all their needs.

Though I have discussed
my differentiating of
brain and mind
with leading neuroscientists
and have received the tentative
approbation of many for my hypothesis,
and have exposed the concepts
to hundreds of audiences
including audiences of teachers
and prominent journalists,
I have had no serious rejections.

When in the spontaneity of a moment
I chose the neurosurgeons
as the particular audience
before whom to differentiate between brain and mind,
I did so
because experience suggested them to be
the most competent judges
of the merit of my theretofore unpublished hypothesis
and method of its evolvement.

Having first resolved that
only the known data
either of inadvertent experience
or deliberate experiment
constitute the foundations
of scientific exploration and discovery,
we will now go on to consider all of such data
that I know to be pertinent
to the potential differentiation
of brain and mind.

For the last three decades
physiologists and neurologists
have been probing the human brain
with electrodes, oscillographs,
potentiometers, et al.
They have combined their instrumental findings
with additional behavioral observations
of first, second and third parties.
The brain probers have identified
several types of energy emanations
both as electromagnetic amperage

and as wave-frequency oscillations
in unique magnitude sequences.

With vast numbers
of permanent bed cases
in veterans' hospitals,
many of those permanent invalids
have willingly and interestedly submitted
to wearing
of the sensationless
head-mounted electrodes
of the brain-probing scientists,
whether they are awake or asleep.

While they sleep
the recording oscillographs
scribe their wavy lines
of magnitudes and frequencies.
These taped records are numbered.
When the patients awake
they are asked to describe their dreams,
if any.

From time to time
the oscillographed wave patterning
is found to be repeating
an earlier recording.
The scientists have found
that unique patterns of waves
characterize specific dreams
which are being re-experienced!

Sum totally to date
the scientists have learned
that the human brain
is a vast communication system
able to record and retrieve information
at varying rates of lag.

The brain is a special case
concept-communicating system
very much like a television set.
It's not just a telegraph wire,
not just a telephone,
it is omnisensorially conceptual as well.
It deals with our optical receipts
as well as with our hearing,
our smelling
and our touching.
In effect we have a telesense station
wherein we receive the live news
and make it into a video-taped documentary.
In our brain studio
we have a myriad of such videoed recordings
of the once live news,
all of which we hold in swiftly retrievable storage.

You are the TV studio's
production director
surrounded by many repeater
cathode-ray tube sets —
you say, "What is going on here?"
as you view — hear, smell, feel — the news.
"Can I recognize this scenario?"

"Have I seen it before —
or anything like it?"
Your phone-headed assistants search the files
and plug in any relevant documentaries.

In any television station studio
the director intuitively sorts and selects
resource sequences
from out of the myriad of
relevant scenarios live or replayed;
now putting the subject at long range,
in full-environmental perspective
and now at close range
scrutinizing some detail.
Other cameras have
their lenses aimed
at static photographs,
others feed in clips
from a host of yesterday's
documentary footage.
The director also has available
imaginarily invented footage
as well as yesterday's experience clues
which may be appropriately considered
at various stages for mixing in with the news.

Out of all this comparative viewing
the director then selects
an appropriate action scenario
of action to be taken now
in view of both the news challenges
and the documentary reminiscences.

Because the brain's TV prime resource
consists of images,
we may call the total brain activity
image-ination.

My youth began a half century before TV.
During that half century
those who wished to discredit
another man's thinking, words or actions
often said,
"Pay no attention to him;
he is full of imagination."
This was tantamount to saying,
"He is a liar."
TV society is not making that mistake any more.

All we have ever seen
is and always will be
in the scopes of our brain's TV station.
All that humanity has ever seen
and will ever see
is its own image-ination;
some of it is faithfully reported new,
some of it is invented fiction or make believe;
some of it is doggedly retained "want to believe."

The physiologists and neurologists
probing the brain
say it is easier to explain
all the data they have
concerning the general phenomena
operative at the top of the spine
if they assume two prime variables always to be operative.

They give one of them the name, "mind,"
the other they call "brain."
The neurologists and physiologists say
it is easier to explain all the data they have
if they assume
the *mind* as well as the *brain* to be *co*-operative,
than it is to explain all the data
if they assume only brain to be operative.

And why is that?
It is because it is found that there are conversations
going on over this communication system —
using its information retrieving and storing system —
whose conversational contents
are in no way explicable
as being produced by feedback of the system itself.

Since neurologists and physiologists
find it desirable
to assume the two phenomena,
the brain and the mind,
I became intent, if possible, to differentiate scientifically
between brain and mind.

I've developed my own
experimental strategy of differentiation
which was published in the 1965 spring edition
of Phi Beta Kappa's quarterly magazine,
The American Scholar.
I have had generally favorable response to it.
I have also the favorable response
of the two thousand neurosurgeons
when I delivered their Harvey Cushing Oration.

The physicists and mathematical physicists call me
an "experimental mathematician"
because I reject axioms and I
explore mathematics experientially.

To present my scientific differentiation
of brain and mind
I proceed experientially as follows —
first I say:
"I take a piece of rope and tense it."
As I purposely tense it
I inadvertently make it tauter.
But I was not tensing the rope
for the purpose of making it tauter,
I was trying only to *elongate* the rope.
Its girth is inadvertently contracting and
the rope is also inadvertently getting harder.

In contracting and getting harder
the rope is contracting in radial compression
in a plane at ninety degrees to the axis
of my consciously purposeful tensing
and — in an inadvertently complementary manner —
next I *purposely* produce compression.
To do so I take tempered-steel rods:
each rod is three feet long
and one eighth of an inch in diameter.
I take one rod by itself
between the fingers of my right and left hands
and press the ends toward one another,
the rod bends flexively.

Using uniform diameter rods,
we find experientially
that two parallel rods cannot stand closer to one another
than in tangency to their circles
of respective cross-sectionalling.
A third parallel rod cannot stand closer
to the other two than by nestling into the valley
between the other two's tangent circles.
With each of the three rods
now in parallel tangency
with both of the other two,
the centers of their three circular cross-sections
form an equiangular triangle.

Hexagons consist of six equiangular triangles.
Hexagons have six circumferential points
and a center point — seven in all —
all equidistant from one another.
Six of our rods now stand together
in closest packed parallel tangency
around the original rod
making seven in all.

Twelve more parallel unidiameter rods
may now be stood in parallel tangency
to form an additionally complete
hexagonal perimeter ring
around the first seven,
making a total of nineteen rods
with each of the interior rods
surrounded tangentially by six others.

Now eighteen more equidiameter rods
standing in parallel tangency
will form an additionally complete hexagonal perimeter
around the first nineteen,
each of which encircled nineteen
will also be nuclear in itself,
this is, be completely surrounded
by six others in closest tangential triangulation.
We may add more parallel tangent hexagonal rings,
each outer ring increasing
the perimeter number by six more rods
than those of the previous outermost ring.
Each of the outermost ring's rods will always
be tangentially closest packed
with only three other rods —
that is, they will be triangularly stabilized,
but not nuclear,
while all the internal rods will be nuclear,
having six tangentially parallel rods around each.
However, only the rod at the center of them all
is the symmetrical nucleus of the whole aggregate.
It is not irrelevent to note
that the rod-like Earthward trajectories
of closely falling inter-mass-attracted raindrops
passing through freezing temperatures
nucleate in hexagonal snow-flake arrays
under just such close-packing laws.
The Greek architects found experientially
that when a stone column's height
exceeds eighteen diameters of its girth
it tends to fail by buckling.
The length to diameter ratio

of compressional columns
is called its *slenderness* ratio.
Steel columns are more stable
than stone columns.
Steel columns are structurally usable
with slenderness ratios as high as thirty-to-one.
But such columns are called long columns.
A short column is one whose slenderness ratio
is far below that of the Greek column.
Short columns tend to fail by crushing
rather than by buckling.
A twelve-to-one slenderness ratio
provides a short column.

For our experiment in *purposeful compression*
we assemble a column thirty-six inches high
with a minimum girth diameter of three inches.
To produce this twelve-to-one short column
we will take 547 of our uniformly dimensioned rods
of one eighth of an inch diameter
by thirty-six inches long.
Each rod by itself has
a slenderness ratio
of two-hundred-and-eighty-eight-to-one,
which is a very highly buckleable column
as we have already discovered.
By compressing it axially,
i.e., bending one end toward the other,
these 547 willow-slender rods
will close-pack symmetrically
into a tangential hexagonal short-column set
of thirteen concentric rings around the nuclear rod.

Standing the hexagonal bundle vertically on a table,
I bind them tightly together
with a steel wire
in a hexagonal "honeycomb" pattern
similar to that of a wire rope cross-section.

The maximum diameter
of this parallel rod packet
will now be three and three-eighths inches.

I next put hexagonal steel caps
neatly fitted over the opposite ends
of the wire-wrapped
steel-rod bunch.
The whole bundle
is now integrated
as a single column
three feet long
and three and three-eighths inches in diameter.

I put this stout short column's ends
between the upper and lower jaws
of a hydraulic press
and thus load the composited rod column
in vertical compression
in the axis of the rods.

We know by our earlier trial
that each end-loaded rod can bend;
so end-loading them in a bunch
results in each rod tending to bend in its middle
but being closest-packed together
they cannot bend inwardly toward one another,

i.e. toward the column's center rod,
they can bend only outwardly away from one another.

Because the binding wire around the rods can stretch,
the binding wire wrapped around the rods yields
to the severe hydraulic loading force
while the bunched ends are held together
by the hexagonal steel caps.
This results in the whole column
becoming cigar-shaped as seen in vertical profile.
If loaded sufficiently,
the bundle approaches sphericity.

This experiment indicates
that our purposeful loading of the column in compression
inadvertently results
in its girth increasing in diameter,
which brings about *tension*
in the horizontally-bound wire —
which is to say, that while I was consciously applying
only a compressive force upon the column's ends,
an inadvertent tension occurs in a plane
at ninety degrees to the axis
of purposeful compressioning.

By two visibly different experiments,
one with rope and one with steel rods —
I have demonstrated experimentally
that tension and compression
always and only coexist.
One can be at ''high tide'' of visibility,
and the other coincidentally
at low-tide visibility.

These always and only coexisting variables,
(where one is at high tide
while the other is at low tide)
are typical complementaries
which are not mirror-images
of one another but must always
complexedly balance one another in physical equations.
Both demonstrate ninety-degree inadvertent resultants.
This behavior is known as the Poisson Effect.

———————————————

Universe is the aggregate
of all of humanity's
consciously apprehended
and communicated experiences,
which aggregate
of only partially overlapping events
is sum-totally a lot of yesterdays
plus an awareness of now.

Yesterdays and now
are neither simultaneous
nor mirror-imaged;
but through them run themes
as overlappingly woven threads,
which, though multipliedly individualized,
sum-totally comprise a scenario.

No single frame either explains
nor foretells the whole continuity —
the picture of the caterpillar
does not foretell the butterfly,

nor does one picture of a butterfly
show that a butterfly flies.
I cannot think simultaneously
about all the special-case events which I have experienced,
but I can think of one special set
of closely associated events
at any one now.

Each one of these thinkable sets
is what I call a *system*.
A system is a subdivision of Universe.
A system subdivides Universe
into all of the Universe events
which are irrelevant to the considered set
because (a) they are outside the system,
too macrocosmic and too infrequent
either to fit into
or to alter
the considered set.
Or irrelevant because (b) occurring
too microcosmically remote
within the system
and of too-high frequency
and of too-short duration
to be tunable with
or to alter significantly
the considered think-set.

After dismissing momentarily
both the macro and micro irrelevancies
there remain
the few clearly relevant sets of associated events
which constitute the system.

We find that all the systems which subdivide Universe
into insideness and outsideness,
are concave on the inside
and convex on the outside.

We next observe that
the concave and the convex
always and only coexist.
A rubber glove on my left hand
has an external part which is convex
and an inside which is concave.
If I strip the rubber glove off my left hand,
it now fits my right hand.
What had been the concave
becomes the convex
and the convex becomes the concave.

So we find these always and only coexisting
complementaries are behaviorally interchangeable —
the tension could be the compression
and the compression could become the tension,
because in one case the girth
went into tension
and in the other it
went into compression.

Please do not think
that we have forgotten
that we are concerned here
with *scientific*, i.e., experimentally based, search
for a means of differentiating neatly
between brain and mind.

Experimental demonstration
of a plurality of special-case instances
of always and only co-occurring phenomena
are prerequisite to generalizing
the brain-mind differentiation.

I will give another example
of always and only co-occurring phenomena.
Physicists today observe
that the proton and neutron
always and only co-occur.
While they are not "mirror" images of one another
and have different weights,
they are transformable
one into the other,
and are thus complexedly complementary,
as are isosceles and scalene triangles.
None of the angles and edges of either need be the same
to produce triangles of equal area.
And the sums of the three angles of each
will always be one hundred and eighty degrees.

The mathematical balancing or complementation
of the proton and neutron are analogously balanced,
each one having two small energy teammates.
The proton has its electron and its antineutrino,
and the neutron has its positron and its neutrino.
And each of these little three-member teams
constitutes what the physicist calls *half-spin* or a *half-quantum*.
They complement one another
and altogether comprise one unit or quantum.

We have now discovered experientially
an always and only coexisting tension and compression;
an always and only coexisting concave and convex;
and an always and only coexisting proton and neutron.
We next consider the *Theory of Functions*
which embraces all of these terms.

X and Y are the always and only co-occurring
covariables of the theory of functions.
We can have X stand abstractly for *tension*
or for *convex* or for *proton*
and we can then have Y stand abstractly
for *compression, concave* or *neutron*, respectively,
in each of our foregoing *always and only co-occurring*
experiential observations of interessential relationships.

We can go further still:
we have the word "relativity."
We cannot have relativity
without at least two phenomena to be differentially related.
There is also the word *complementarity*.
We cannot have one phenomenon complemented
by less than one other phenomenon.
The words *complementarity* and *relativity*
do not identify identical physical phenomena.
We need to discover
whether there exists a generalized concept
which embraces both phenomena,
and we find that the ponderable physical energy Universe
that is, *physical universe,*
in contradistinction to the Universe's
weightless, metaphysical aspects,
does embrace both *complementarity* and *relativity*.

I started our brain-mind differentiation
by saying, "I take a piece of rope."
I've done this before many audiences,
and no audience has ever said,
"You don't have a piece of rope."
But the fact is I *didn't* have a piece of rope.
Nor has anybody ever said,
"Is it nylon, manila or cotton?"
or, "What is its diameter?"

My statement was a "First-degree scientific generalization."
In literature the word *generalization* means
covering too much territory
too thinly to be convincing.
However, we have in science a term "generalization,"
which does not have the literary connotation.
A *generalization* in *science* refers to
a principle discovered by experiment
to be operative in every special case.

If we find any exception,
we no longer have a scientific generalization.
Scientific generalizations are extraordinarily meaningful,
as for instance was the discovery
of the principle of leverage,
which probably came about as follows:
occasionally humans who have penetrated
wilderness forests
encounter trees fallen slantwise
across their line-of-sight path
in their chosen direction of travel.

It is obviously quicker
to climb over the fallen tree
than to try to walk around it.
They find it logical
to walk along the top of the fallen tree
when it leads in the direction of preferred travel
or toward the next opening in the forest.
As they walk along the horizontal trunk
they feel the tree to be progressively sinking.
As they move farther it tips earthward
at a faster rate.
They retreat —
back along the tree trunk.
Experimenting, they find the tree
on which they are walking
is lying across another tree.
And then they observe
that the end of the tree behind them,
opposite to the direction
in which they were walking,
is itself superimposed
by a third and very mighty tree.

Looking the situation over they find
that as they walk outward — journeyward —
along the first tree
that its slow but accelerating descent
coincides with the other end's
lifting the trunk of the massive tree.
Never having heard of a lever
or fulcrum,
they say, "That big tree which is being lifted
is much too big for me to lift."

They go over to the massive tree
and attempt to lift it directly
with their arm, back and leg muscles.
It doesn't budge.

Shaking their heads in surprise,
they once more try walking along
on the first tree.
Again the massive tree rises easily.
And Neanderthal man probably thought
as it rose
that he had found a magic *tree-lifting* tree.
And he probably dragged it home
where the tribe worshipped it
until suddenly his wife said,
"Any tree will do that lifting."
And sure enough,
not only would any tree do
but so too would any steel bar,
or glass reinforced plastic bar,
or any small-toothed gear,
or a large-toothed gear.

This man discovered
a true scientific generalization
which always holds true
under any circumstances.
The lever works equally well
anywhere in Universe.
It can be made of many materials,
it can be of any size.
Its behavior follows incisively predictable
mathematical laws.

It was a first-degree generalization
when I said, "I take a piece of rope,"
and in describing my purposeful tensing of it,
there was nothing I said
regarding the piece of rope
that in any way contradicted
any experience that anybody
in any audience has ever had
with any piece of rope.

It is a second-degree generalization
to find an additional generalized principle
operating within the generalized piece of rope,
such as the always and only
coexisting tension and compression.
It is also a second-degree generalization
to find the concave and convex within generalized systems.

It's a third-degree generalization
or a generalization of a group of generalizations,
to develop the Theory of Functions
wherein the X and Y could stand
for any second-degree coexistence generalizations.

It is a fourth-degree generalization
to develop the word "relativity."

And it's a fifth-degree generalization
to employ the word "Universe"
to embrace both the relativity and complementarity.
The degrees are then progressive omnibus stages
of generalizations of generalizations.

The generalized principles
are all interaccommodative.
None contradicts another;
nor do they contradict
any combination
of other scientific generalizations.
They have complete integrity.
One is associable with the other.

The a priori integrity
of scientific generalizations
is manifest in all the phases of Universe
as we explore it progressively.
Humans do not create,
humans cannot create.
Creation is a priori,
creation is the gamut
of generalized principles
which scientists can and do discover.
And inventors can employ
in special-case uses.

Humans can invent,
which means "bring in,"
the special-case use
of generalized principles
and of combinations of them.
But humans cannot design
or invent
a generalized anything.

We play tension and compression with a little dog.
He uses compression of his teeth

as we pull on the stick and tense it,
and the dog uses
the convex and concave surfaces of his teeth
and his protons and neutrons
are in beautiful co-ordination
without his even knowing it.
But there is no experience with dogs
that suggests that they could ever
develop the Theory of Functions.

Some living creatures' brains
develop conditioned reflexes
to special-case type events,
which produce behaviors resembling a first-degree
sense of generalization.
But there seems to be no indication
that they ever evolve
a generalization of a generalization,
and they certainly do not get to
fourth- and fifth-degree generalizations.

We find that brains deal
always and only with special cases.
We remember the special-case name — Tim Smith —
which we may not need to recall
for thirty-five years,
yet it occupies its own special
physical brain's "after-image" locale —
possibly its own neuron.

We find the brain always dealing with special cases
and the mind dealing with generalizations.
The generalizations have no weight;

only special cases have weight
or any other physical characteristic.
All scientific generalizations
are pure metaphysics.
All metaphysics are weightless
and physically unlimited.
A triangle is a generalized principle
and a triangle persists conceptually
as a triangle,
independently of size,
color, weight, taste, texture or time.

Searching for a function of humans in Universe
we found first
a phase of Universe
which needed counterbalancing.
This was the entropic, expansive,
increasingly disorderly, radiantly explosive Universe.
In answer we found Earth to be a space sphere
wherein Universe was collecting
sorting, concentrating, and storing energies
in mathematically orderly ways
such as by refraction, crystal growth,
photosynthesis or molecular formations.
Then we found that all the biologicals
were antientropic.

I found myself publishing in '49
as did Norbert Weiner in another book at the same time,
unwitting of one another's coincidental
perceptionings and simultaneous disclosures
that all biological life is antientropic
and that the human mind

is the most powerfully effective
antientropy
thus far evidenced.

It was only a few years ago
that it seemed logical
to cease speaking of the phenomena involved
as antientropy,
entropy being disintegratively negative;
antientropy was, in effect, a double negative
used to express a positive.
So to render the concept positive
and to identify its kinship
to synergy,
I started speaking of it as *syntropy*,
as the positive complementary
of the negative entropy.

Now I surmise
that the speculative thought
of the human mind —
in contradistinction
to the physical experience recalls
of the physical brain —
is physically nondemonstrable,
ergo, metaphysical,
but its teleological activity,
which subjectively evolves generalizations
from multiplicities of special-case experiences
(and thereafter employs the generalizations
objectively in other special-case physical formulations)
can be detected
through our intuitive recognition

of the weightless pattern integrity per se,
ever weightlessly, abstractly present in the original
discovering and inventing events.
Scientific generalizations
have no inherent beginning or ending.
In discovering them
mind is discovering a phase of Universe
that is eternal.
The physical human and its physical brain
unmonitored by mind
are not only less effectively syntropic
than other biological species;
they are often consciously
far more entropic
than any other species
and are only subconsciously syntropic.

Human's average birth weight of seven pounds
is multiplied sometimes as much as fiftyfold
by syntropic chemical combining of
other protoplasmic cells
taken in as food,
and with gases and liquids
taken in by breathing and drinking.
But humanity's syntropic cell multiplication
is outperformed greatly by trees, whales, elephants
and other living organisms.

Remembering that antientropy, or syntropy
means, on the one hand,
reversing the physical trending
toward greater disorder
and, on the other hand, means

converting the relative disorder
to increasingly orderly physical arrangements
we found that the human's biological role is minor,
but that the human's metaphysically-sorting mind-role
is Earth's paramount syntropic factor.
For it reviews and sets in fundamental order
the randomly, ergo disorderly,
physical event experiences of the universal environment.
And the magnitude of physical rearranging
in progressively more orderly patterns
metaphysically conceived by the human mind
is overwhelmingly greater
than that produced by the
physiological organisms
of any other biological species.
One mind can design orderly machines,
such as the steamship *Queen Mary,*
or the Great Wall of China
in physical magnitudes which dwarf
any biological undertaking,
even that of coral-reef building
by minuscule coral creatures.

It must be remembered
that syntropy also means
to collect, concentrate, and store.
When I speak to an audience
as recounted before,
saying, ''Let us take a piece of rope...''
to demonstrate the generalized rope concept —
I am drawing on
a multiplicity of special-case rope experiences
as a brain-stored resource of that audience,

probably amounting to over a hundred experiences each
with different kinds of pieces of rope:
ergo — I am drawing upon a memory resource
of more than one hundred thousand experiences
with as many different pieces of rope,
when I speak to an audience of one thousand.

So my first-degree generalization
reduces our experience processing
one hundred thousandfold.
Our second-degree generalization
reduces the experiences multi-billionfold,
and each further degree of generalization
multiplies the number of special experiences
from which the generalization is extracted.

The third-degree generalization,
consisting of the theory of functions,
embraced all the always-and-only co-occurring phenomena
of a plurality of second-degree generalizations.
When we get to the fifth-degree generalization,
Universe,
we have increased our number-of-experiences base
to all the experiences
ever known to and remembered by humanity,
including all the experiences
with all the atoms and their nuclear components.
Thus, the human mind
has collected, combined and refined
all experiences of all humanity,
in all-remembered time,
into one single concept,
Universe,

which is, *ipso facto,*
the ultimate generalization.

Clearly it is seen
that the human metaphysical mind
demonstrates the most effective
syntropic capability evidenced in Universe
excepting that of the universal mind's
cosmic syntropy
as manifest in the *eternal design complex,*
comprehensively and synergetically interaccommodative
as eternally regenerative Universe,
which cosmic syntropy combines
all the metaphysical integrities
as well as all the physical patterns.

Human mind's metaphysical syntropy potential
caps all the syntropic sequences
operative in our Spaceship Earth's
comprehensive syntropic system,
which, as we remember, started its analysis
hoping to be able
to identify humanity's function in Universe
and thereby to gain insights
into the respective functions
of the mind and brain.
Reviewing our strategy once more
for synergetic feedback we recall
that we started looking for
the syntropic complementation
of the entropic energy expenditures
of all the stars
and found our own Spaceship Earth

to be the most immediate demonstration
of the energy aggregating
of star radiation and stardust,
and we observed
the progressive impoundments by the atmosphere
and oceans, of energy as heat,
refracted into circumferential
weather and ocean currents,
which energy impoundments produced the exact
temperatures, pressures, and chemical wherewithal
for developing the vegetational photosynthesis
of land and sea.

And we went on to find that
the effect of all of this is
to reduce the randomly received energies
into superbly orderly
molecular and cellular structurings
and to multiply that highly concentrated
and chemically locked-in
energy — as hydrocarbons
which were and are continually
being buried deeper and deeper
within Earth's agglomerate surface,
where the pressures and heat
at a kilometer's depth
begin to convert the hydrocarbon fossils into petroleum
as an even more compacted energy conserve.

And beyond these physiological syntropies
we find the human mind
discovering the generalized principles
operative weightlessly and eternally

everywhere and everywhen
in Universe,
knowledge of which principles
permits humans to employ
the vast extraterrestrial energies
to do evermore rigorously scientific,
evermore efficient rearranging
of the random terrestrial environment;
thereby to regenerate and sustain
evermore years of evermore humanity's lives.
We saw that humans will learn how to concentrate
and compress evermore minuscule packages of energy
to impel human travelers —
so that when the time comes,
millions of years hence,
for the Earth-concentrated energies
to become an energy reradiating star,
the humans will have migrated
safely elsewhere in Universe
to perpetuate its supreme syntropic functioning
in Scenario Universe
as the ultimate sorting,
rearranging, compacting
and logic-employing local monitor
of the syntropic phases of regenerative Universe.

Einstein,
as metaphysical, weightless, human intellect,
took the measure of the weighable physical
and wrote
the most economically formulated equation
mathematically possible,
for it put on one side of the equation

the physical Universe, which is to be equated,
represented as "E" —
because all that is physical is energetic —
and on the other side of the equation
Einstein placed the two terms
minimally necessary to disclose a relationship.
For these two terms he employed
"M" (for energy associated as matter) and
"c^2" (for energy disassociative as radiation),
expressed as c^2 because
c is the speed of light
in any one linear direction;
but light goes omnioutwardly in a spherical wave
whose surface increases as the second power
of the linear speed of radiation.

Ergo: Einstein's equation
is $E = Mc^2$.
And in it the velocity of c^2
is the constant and known term
and the amount of energy
compacted in any given mass
can be determined
if the given E to be analyzed
is measured.

Einstein's equation
was intuitively formulated
from the experimentally
harvested data of others,
but proved to be correct
when the subsequent fission occurred.

Here we have the metaphysical intellect
taking the measure of
and mastering the physical.
We have no experimental data
that in any way suggests
that this process is reversible
and that energy can or will
take the measure of
and write the equation of intellect
or the equation of the metaphysical.

This is an example of one
of the great generalized principles
operative in Scenario Universe,
which is the principle
or irreversibility
of intellectual processes.
For the syntropic metaphysical
is not a mirror-imaged reversal
of the entropic physical's
disorderly expansiveness.

The fact that humans,
using only physical brain
and not mind,
can be the most
entropically destructive organism
does not contradict
the irreversibility principle
unique to maximally syntropic mind.
Humanity's imaginative invention of Hell
discloses its subconscious awareness
of the ultimate entropy.

Thus we find the metaphysical
apprehending and embracing,
comprehending, cohering and conserving
the integrity of Scenario Universe's
never exactly identical recyclings.
The physical tries to destroy
and dissipate itself.
The metaphysical law masters and conserves
the evolutionary integrity.

Though humans are born equipped
to participate
in the supreme function of Universe,
this does not guarantee
that they will do so.
Humans are born utterly helpless,
and must through trial and error,
by physical experiments,
discover what the controlling family
of generalized principles may be,
which principles must be employed by humans
to fulfill their Universe function.
But in order to discover the latter
they first must discover the overwhelmingly superior efficacy
of the mind, as compared to muscle.
And humans also must discover
that the physical,
which they tend to prize as seemingly vital
to their regenerative continuance,
is utterly subordinate
to the omni-integrity of metaphysical laws,
which are discoverable
only by mind.

Only if humans learn in time
to accredit the weightless thinking
over the physical values,
in a realistic, economic and philosophic accounting
of all their affairs,
will the particular team of humans
now aboard planet Earth
survive to perform their function.

When nature has an *essential*
and intercomplementary function to be fulfilled
and the chances of development
of that functioning capability
are poor
nature makes many potential "starts."
As for instance on planet Earth
all the vegetation which impounds the Sun's energy
must be regenerated and multiplied,
but it cannot have its progeny
within its immediate vicinity,
as the tree's shadow
would prevent its young
from further impounding the Sun's radiant energy.

Wherefore all the trees
launch their seeds
into the air or upon the waters
to drift to chance landings,
where the seeds may be favorably nourished
and grow.
The chances of such auspicious landing
are so unfavorable
that nature must send

billions times billions of seeds away
from the parent vegetation,
which, though potential of complete success,
may never germinate and prosper.
The airs and waters
of the planet Earth
are filled with the aimlessly migrating seeds.

Because the chances of humanity's
self-discovery of the supremacy of the metaphysical
and the corruptibility of the physical,
while coming from an utterly helpless start,
are very poor,
the probability is
that for each of the billions of stars
in the billions of nebulae
there are several planets
where energies are being
most effectively conserved —
which means
by the metaphysical mind.

Ergo: there are probably myriads
of successful human mind counterparts
on consciously operated planets
despite greater myriads of failures.

One physicist remarked recently,
"I am tiring of the nonsense legend
which finds one end of Universe closed,
by a required beginning event
and the other end open to infinity."
The concept of primordial —
meaning before the days of order —
which imply an a priori,
absolute disorder, chaos, *a beginning*
("the primordial ooze-gooze explosion")
is now scientifically invalidated, passé.

The physicist finds
that the proton and neutron
not only always and only co-occur,
and are interchangeably transformable,
but also could not occur independently
anymore than a triangle could occur
with only two points.

We cannot have disorder
because Universe is not monological;
it is *pluralistic* and *complementary*,
and we are founded on
the orderly base
of the proton-neutron tripartite teams
of six unique energy integrity vectors.

Men at the time of World War I
were dealing in radio waves
that were a mile long.
Gradually they found shorter waves
and found ways of sending,

transmitting and receiving
at ever shorter and shorter wavelengths
and at higher and higher frequencies.

A quarter of a century later,
when we came into World War II
the electronics scientists
were working at wavelengths
of about two meters.
After World War II they were working
with what we call *micro*waves,
operating at fractions of meters
and at fractions of centimeters or millimeters
of wavelengths.
The higher the frequency
the shorter the wavelength —
and the more tendency
to interference
with the waves of other phenomena.
Also: the higher the frequency,
the lower the energy required
to power the propagation
of the message-carrying
electromagnetic waves.

Therefore, these short, high frequencies
could only be sent
relatively short distances.
They are interrupted by walls, hills or trees.
The long waves are not interfered with
by such local obstacles.

Those who employ
very high frequency short waves find
their short waves are so interrupted
by local events
that the walkie-talkies, for instance,
cannot carry far enough
to interfere with other stations such as
the commercial or government traffic,
because they do not interfere
with longer waves of the commercial bands.

It is also a characteristic of these waves
and of all radiation
that when the wave propagation
is beamingly aimed
perpendicularly outward from Earth's surface
they experience little or no interference,
once outside our atmo-, strato-, and ionospheres,
other than by collision with meteorites
and other celestially traveling objects.

There seems to be no impedance
and no inherent limitation to the distance
which such electromagnetic wave signals can go
once outside the Earth mantles.
As far as we know,
the waves can go on forever in Universe —
unless they hit some object,
and when they hit an object they lose some energy
then bounce away
and keep on going
in a new direction.

Now let us turn our thoughts to the neurologists
who we left probing the brain with electrodes.
I find that the neurologists do not feel it to be shocking
when I suggest to them
that it could be demonstrated physiologically
within the next two decades
that what man in the past has been calling telepathy
may in fact be ultra, ultra-high frequency,
electromagnetic-wave propagation.

Almost everyone has had
the strange sensation of telepathy
occurring as various kinds of awareness,
anticipations or sensing
of the imminent presence of other persons.

I myself had a very extraordinary experience
with our first child,
who died just before her fourth birthday.
Born at the time of World War I,
she first contracted flu,
then spinal meningitis,
and finally, infantile paralysis,
being paralyzed and unable to move,
as do other children,
but with her brain unimpaired.
She manifested the normal child's innate drive
to apprehend then comprehend
all that her eyes could see,
her nose smell,
and her ears hear.
She had to obtain
the tactile information

by other means
than with her own hands;
so the tensive-compressive,
soft, hard, light, heavy,
rough, smooth, wet, dry,
hot, cold, cool and warm characteristics
of the constituents of her environment
she had to sense vicariously.
And in so doing,
she demonstrated the extraordinary
compensating faculties
innate in all humanity,
which are, however,
brought into use
only under very special conditions.

She was fantastically sensitive
to the cerebrations of all humans in her vicinity.
She was most frequently attended
by my wife, myself
and two trained nurses.
Often she had two of us with her.
Time and again when we were
about to say something to one another
which was not the kind of concept
that would be known to a child of her age,
she would say what we were about to say
before we had time to say it.
This happened so many times
as to convince us that our formulated communications
were obviously being transmitted through her.

This experience persuaded me that telepathy
might well be very short-range,

very high-frequency
electromagnetic-wave propagation.
Assuming this to be so
we arrive at some new vistas of thought.

When we *broadcast* energies
they are very greatly dissipated.
Radiant energies can be concentrated, however,
by reflective beaming and lensing,
as was candlelight in a lighthouse.
Reflectors and lenses concentrated them.
Reflectively beamed seaward,
they were sometimes
visible for ten miles.

After World War I the only unused
electromagnetic-wave-band frequencies available
were by international conventions preassigned
to the then developing,
but not as yet inaugurated
prime television-developing countries.
These available bands
were in the very short-wave
and high-frequency area,
where interferences were so frequent and great
that beyond-the-horizon broadcasting
was physically impossible.

Unfamiliar as yet with the beaming technique
and its more economic potentials,
it was the accepted professional
engineering dictum
that if we were going to have
worthwhile, desirable, and popularly sustained —

and therefore commercially exploitable —
post World War II TV programs,
we would not be able to afford
having top-rank artists
and high-cost programs
moving to every little local center
to be locally broadcast.
Therefore we would have to have
all expensive TV programs
developed at central places
and distributed by cable
to secondary broadcasting stations.

During World War II
the discovery of the
shorter and shorter waves and the capability
to receive and transmit them
brought about after the war
the facile use of short waves,
which waves were found to be short enough
to be easily reflected.
If the wave is a mile
in length, we have to have
a reflector over a mile in size.
As with optical light reflection,
we have to contain the wavelength
in order to beam it.

Wherefore after World War II
when television was instituted
we had masts at horizon-to-horizon points,
well above the interference patterns
of trees, mountains and buildings.

Instead of sending the radiation in all directions
we concentrated it in one direction,
which greatly conserved it.
And energy boosted the signals
at each horizon relaying transceiver,
reconcentrated them,
and sent them onto the next horizon point.
In this manner the TV programs
are now transmitted around the world's local national domains
by satellite-relay transceiver.

Recognizing that it is possible
to conserve energies by reflection,
as well as to reach
great distance by beaming,
we can point out that it is also possible
that our human eyes are just such
very high-frequency electromagnetic wave
propagating and receiving reflector relays,
as with the original propagation
occurring in the brain
and the transceiving
relayed by our eyes.

I have had extraordinary experiences
with audience after audience
around the world.
I find that eyes tell me so much
that I am able to go into a room
wherein some verdict has been adopted,
and I find that
I know what the verdict is
before anyone has spoken
audible words.

And I am confident
that I first "saw" the message
in the people's eyes
and not in their facial expressions,
which were mixed and arbitrarily fixed.
The speed of light,
ergo, of the sight functioning,
which is approximately
seven hundred million miles per hour,
is such an enormous velocity
that we mistakenly sense it
only as "instantaneous."

When I was young
a camera required a minute's exposure.
By improving film chemistry
and lens refinements
as we entered World War II,
a thousandth-of-a-second exposure
had become adequate
for premium photographic equipment.
Now *adequate* exposure
with some scientific equipment
has been lessened
to one millionth of a second
for producing a superb photograph.
In other words, the rate
at which we can "get the picture,"
and the rate at which we can transmit it
is approximately instantaneous.

It may be true
that our eyes are electromagnetic-wave transceiving relays.

If so, it is possible
that seemingly instant exchanges between humans
may become scientifically accomplished through telepathy,
and people may soon be able
to know one another's thoughts,
which will mean
that people will be prone
to do good thinking
and also to co-ordinate
with one another
as never before.

In this connection it is recalled that when
Artzybasheff made a picture of me
for the cover of *Time* magazine,
he pictured my head as a geodesic sphere.
The shape of my head is
of course not that of an exact sphere.
Everything about my features
except my eyes
was purely mechanistic,
and were caricatured as mechanical devices.
Though I sometimes read of myself
as being almost inhumanly mechanistic
I myself think of the whole physical Universe
as governed so exquisitely by generalized laws
as to have to conclude
the whole of physical Universe
is technologically governed,
but I find life to be weightless
and only metaphysical
and I know that I am inspired
entirely by life

and its needs and potentials.
So I do not tend to think
such characterizations by an illustrator
to be accurate.
Yet this *Time* cover picture seems to me
to be the best portrait of me
that I recall having seen.
I never met Artzybasheff.
He is now dead.
But he sent word to me
before his death,
relating his special satisfaction
with this particular portrait of me.
He also sent word
that the only features that mattered to him
were the human's eyes.
Before he painted my portrait
Time-Life photographers
took many pictures,
but only of my eyes.
Artzybasheff did my portrait
entirely from those eye photographs,
and all the foregoing
discussion of the transceiving,
beam-relaying functioning of the eyes
may explain why
it became the optimum likeness of me
despite the mechanistic caricaturing.

Now let us employ
all the foregoing discussion of wave phenomena
and speculate as to its significance —

and remembering that, lacking interference,
electromagnetic waves
apparently can travel on
at their seven-hundred-million-miles-per-hour rate
for what may be limitless periods of time.
Let us assume a cloudless night
somewhere around the spherical surface
aboard our spacevehicle Earth,
and a human looking out at the stars
and inspired by the celestial splendor
to be thinking profoundly.
It is quite plausible
that his skyward focused eyes
may beam his thoughts,
quite unbeknownst to him,
out through the shallow atmosphere
into the approximately interference-free
macrocosmos.

And there is no reason why
the eye-beamed thoughts
might not someday
bounce off some other celestial body,
as humans have already
bounced radio signals off the Moon
and back to Earth
by carefully angled beaming.

And as at present,
to avoid the circumferential
obstacles of our planet,
electromagnetic-wave-carried programs

of TV and voice
are being relay-bounced
around Earth by the
communication satellites,
holding their flight positions
outside the atmosphere.
And the thoughts of our specially assumed human
who is inadvertently beaming
his thoughts into the cosmos
sends them on a path
which results in their
being uninterfered-with for millennia.

When finally they do interfere with
and bounce off a celestial body,
they are accidentally aimed back to where
Earth will be several millennia after their original dispatch.
At that moment of rereaching Earth planet
an individual on board Earth
is looking out at the night heavens
and inadvertently tunes in the millenia-ago
telepathy-dispatched thoughts
through his transceiver beam-relaying eyes,
and the thought message is monitored into his brain,
whereby the inadvertently receiving human
thinks he is thinking
a novel and interesting thought —
and all of the foregoing
seems to indicate the possibility
that the family
of generalized principles
being eternally valid

independently of special-case idiosyncrasies
(ergo, of the language of its thinker)
might be bouncing around in Universe
to be tuned-in
here and there from time to time
on various planets
by various humans
of various planetary crews.

And it may be thus
that knowledge becomes tune-in-able
by humans on planets or wherever they may be.
This telepathic tuneability may occur
as the humans complete
enough experiences
and do enough generalized thinking about them
to be able intuitively
to comprehend the significance
of the thoughts which they are
inadvertently receiving.

Certain it is on my own part
that I have made several mathematical discoveries
of fundamentally unexpected and unpublished nature.
As I realized my discovery
I always have had
the same strange sensation
that this newly realized conception,
previously unknown to terrestrial humans,
had been known
to the human mind
sometime vastly long ago.

Since whatever life may be,
it has no weight,
as has been discovered
by weighing individuals
at the moment of their dying,
it is also possible,
that whatever our abstract
metaphysical beings may be,
their complex weightless organic pattern integrity
might also be transmittable whole
or by installments of electromagnetic waves,
whereby humans may already have been,
or Earthians may sometime become
beamed consciously and purposefully
to elsewhere in Universe
traveling at seven hundred million miles per hour,
rather than at the ponderously slow rate
of twenty thousand miles per hour to which
our present Earthian rocketing is confined.

And wherever they came from,
the thoughts arranged in this book
are discoveries
of its author
since he first came in 1913
to think
that nature did not have
separate departments of
mathematics, physics,
chemistry, biology,
history and languages,

which would require
department head meetings
to decide what to do
whenever a boy threw
a stone in the water,
with the complex of consequences
crossing all departmental lines.
Ergo, I came to think that nature
has only one department —
and I set to discover its
obviously
omnirational
comprehensively co-ordinate system,
and thankfully found it.

LOVE

TWO VERSIONS OF
THE LORD'S PRAYER

Love

Love
is omni-inclusive,
progressively exquisite,
understanding and tender
and compassionately attuned
to other than self.

Macrocosmically speaking
experience teaches
both the fading away
of remote yesterdays
and the unseeability
of far forward events.
Microcosmically speaking
science has proven
the absolutely exact
also to be
humanly unreachable,
for all acts of measuring
alter that which is measured.

Conceptual totality
is inherently prohibited.
But exactitude can be bettered
and measurement refined
by progressively reducing
residual errors,
thereby disclosing
the directions of truths
ever progressing
toward the eternally exact
utter perfection,
complete understanding,
absolute wisdom,
unattainable by humans
but affirming God
omnipermeative,
omniregenerative,
all incorruptible
as infinitely inclusive
exquisite love.

While humans may never
know God directly
they may have and do
palpitatingly hover
now towards, now away.
And some in totality
come closer to God.

And whole ages of peoples
in various places
leave average records

of relative proximities
attained toward perfection.

PERSIA — positioned
at demographical center
of all Earthian peoples —
has been traversed by many
into and beyond
the vanishing past
and will be traversed by many
into and beyond
for foreseeable future.

And at this most crisscrossed
crossroads of history
the record is left
of the relative proximity
averagingly attained
to that which is God.

The PERSIANS' record
is tender and poignant
sheltering, embracing,
an omnipoetical
proximity to God.

Garden,
Shah Abbas Inn
Isfahan
September 6, 1970

Two Versions of the Lord's Prayer

I feel intuitively that what is now identified as the Lord's Prayer was digested through ages from many philosophies in many lands. Also I feel intuitively that in relaying the Lord's Prayer from country to country, from language to language, from one historical period to another that many at first small, then later large alterations of meaning may well have occurred. It seems unlikely to me that the prayer's original conceivers and formulators would have included a bargaining proposal such as asking forgiveness of our trespasses or debts because we agree to forgive others. It also seems illogical to remind God of anything or to ask special dispensation for self, or to suggest that God doesn't understand various problems, or that God needs earthly salesmen for his cause. Before going to sleep, even for short naps, I always re-explore and rethink my way through the Lord's Prayer. And as I thought it through tonight, August thirteenth, 1966, I decided to inscribe it on paper.

Oh god, our father —
our furtherer
our evolutionary integrity unfolder
who are in heaven —
who art in he-even
who is in everyone
hallowed (halo-ed)
be thy name
(be thine identity)

halo-ed —
    the circumferential radiance
    the omnidirectional aura of
    our awareness of being
    ever in the presence
    of that which is greater,
    more exquisite
    and more enduring
    than self —
haloed be thine identification
which is to say
the omnidirectional
vision and total awareness
is a manifest of your identity
which name or identity is
most economically stated as:
truth is your identity —
    by truth we mean
    the ever more
    inclusive and incisive
    comprehension
    which never reaches
    but always approaches
    closer to
    perfection of understanding
    and our awareness is ever
    a challenge by truth —
truth is embraced by and permeates
the omnidirectionally witnessible integrity
of omni-intertransforming events
which ever transpire — radiationally
which means, entropically in the physical;
and contractively — gravitationally —
which means, syntropically in the metaphysical

both of which — are characterized by either
the physical expansions
toward ever-increasing disorder,
of the entropic physical;
or the metaphysical contractions
toward ever-increasing order
of the syntropic metaphysical;
and these pulsating
contractions-expansions
altogether propagate
the wave and counterwave oscillations
of the electromagnetic spectrum's
complex integration —
of the omnienvironment's
evolutionary reality
and its concomitant
thought regeneration
which altogether constitute
what we mean
by the total *being*.

Hallowed
or haloed
"be thy name," which means
your identification to us,
your kingdom come! —
your mastery of both
the physical and metaphysical Universe
emerges as the total reality
your will be done —
your will of orderly
consideration and mastery
of the disorderly be done
on our specialized case planet

and in our specialized case beings
and in our special-case consciousness
and in our special-case intellectual integrities
and in our special-case teleologic integrities
as it is in your generalized case he-even (heaven).

We welcome each day our daily evolution
and we forgive, post give, and give
all those who seemingly
trespass against us
for we have learned retrospectively
and repeatedly
that the seeming trespasses
are in fact the feedback of our own negatives,
realistic recognition of which
may eliminate those negatives.

For yours, dear god —
oh truthful thought
is our experience proven manifest
of your complete knowledge,
your complete understanding,
your complete love and compassion,
your complete forgiveness —
subjective and objective,
your complete inspiration and vision giving,
and your complete evolutionary volition,
capability, will, power, initiative
timing and realization —
for yours is the glory —
because you *are* the integrity
forever
and forever
amen.

This is the way I thought through The Lord's Prayer on June 30, 1971, at the American Academy in Rome.

Oh god
our father
who art in he even
omniexperience
is your identity.
You have given us
o'erwhelmingly manifestation
of your complete knowledge,
your complete comprehension,
your complete wisdom,
your complete concern,
your complete competence,
your complete effectiveness,
your complete love and compassion,
your complete forgiveness, giveness, and postgiveness,
your complete inspiration giving,
your complete evolutionary sagacity,
your complete power, will, initiative,
your absolute timing of all realization.
Yours, dear god,
is the only and the complete glory!
You are the universal integrity
the eternal integrity is you.
We thank you with all our hearts,
souls and mind —
Amen.